Longman Chemistry homework for Edexcel IGCSE

Martin Stirrup

BRADFIELD COLLEGE

FORM	NAME AND HOUSE (PLEASE PRINT)	DATE BORROWED	DATE RETURNED
11	Freddie McCleon	3/10/07	23/5/08

Pearson Education
Edinburgh Gate
Harlow
Essex
CM20 2JE

and Associated Companies throughout the world

www.longman.co.uk

© Pearson Education Limited 2007

The rights of Martin Stirrup to be identified as the author of this work have been asserted by him in accordance with the Copyright, Designs and Patents Act, 1988.

All rights reserved. No part of this publication may be reproduced, stored in a retrieval system, or transmitted in any form or by any means, electronic, mechanical, photocopying, recording, or otherwise without prior written permission of the Publishers or a licence permitting restricted copying in the United Kingdom issued by the Copyright Licensing Agency Ltd, Saffron House, 6–10 Kirby Street, London, EC1N 8TS.

ISBN: 978 1 4058 7494 6

Development and editorial by Sue Kearsey
Designed by Redmoor Design, Tavistock, Devon
Cover photo © www.istockphoto.com
Printed in Great Britain by Henry Ling Ltd., at the Dorset Press, Dorchester, Dorset

Contents

Contents

How to use this book

This homework book is designed to help you practise all the science you need for your GCSE. The questions are arranged to match the chapters in your Student's Book, as shown in the contents list for this book.

The questions will help you to:

- develop your ideas about each topic
- make key notes or diagrams to use when you revise
- practise solving science problems
- get information from tables, charts and graphs
- see how science affects you and your environment.

The questions are graded on each page, starting with simple exercises and getting harder. Higher tier material is clearly marked with square question numbers. Your teacher will tell you which questions to try.

Most of the information you need will be on the page with the questions, including formulae for calculations. Anything else you need will be at the back of the book. The questions are to help you learn, not to try to catch you out.

Remember Don't be content with just writing down an answer. Think carefully – does your answer make sense? Could you explain it to somebody else? As you do each question, you should: read, think, do, check – and finally understand! If you are still in doubt, ask your friends or your teacher, but work through the answer to make sure you really understand how to get there yourself.

1 Safety in the lab

❶ Copy and complete the table, showing the names of the pieces of apparatus and what they are used for.

Apparatus	Name	Use
	a	b
	c	d
	e	f
	g	h
	i	j
	k	l

❷ Copy and complete the table to show the meaning of each of these symbols. Choose an example of a chemical of each type from this list.

- lead(II) nitrate
- petrol
- copper(II) oxide
- hydrogen peroxide
- mercury
- hydrochloric acid

Hazard symbol	Meaning	Example
	a	b
	c	d
	e	f
	g	h
	i	j
h	k	l

❸ Hazard symbols are found in many places outside laboratories, including in homes, garden sheds, in shops and on vehicles. Copy the table and fill in as many examples of hazard symbols as you can find around your own home. Sometimes the symbols are obvious, sometimes you have to look underneath the bottle or packet to find them. Ask permission before you go searching through the cupboards at home.

Hazard symbol	Where found

2 Inside the atom

1 Copy and complete these sentences. Use the words below to fill the gaps.
electrons nucleus protons subatomic

Atoms are made from even smaller .... particles. In the centre is the ...., which contains the .... and neutrons. The shape of the atom is given by tiny particles called ...., which whiz around the nucleus.

2 **a** Copy and complete this sentence using the correct word from each pair. You should be able to make *two* correct sentences.

Protons/electrons have a relatively **large/small** mass and a **negative/positive** charge.

b What is the charge on a neutron? (Think!)
c Which two kinds of particle are found in the nucleus?
d Draw a simple diagram of a section through an atom, showing the nucleus, protons, neutrons and electrons.

3 The mass of a proton is too small to measure in grams. Instead the mass is compared to a hydrogen atom, which is given the value 1. Copy and complete the table, showing the relative mass and charge of the subatomic particles.

	Mass	Charge
proton	**a**	positive
neutron	1	**b**
electron	0	**c**

4 Copy and complete these sentences choosing the correct word from each pair.

Atoms of the same element have the same number of **protons/neutrons**. The number of **protons/neutrons** in an atom is called its **atomic/mass** number. The number of protons plus the numbers of **electrons/neutrons** gives the **atomic/mass** number. As an atom is neutral, the number of negative **electrons/neutrons** is always the same as the number of positive protons.

5 For each of the following elements, give the atomic number (proton number) and the mass number.
a helium (He) has 2 protons, 2 neutrons and 2 electrons
b fluorine (F) has 9 protons, 10 neutrons and 9 electrons
c iron (Fe) has 26 protons, 30 neutrons and 26 electrons
d uranium (U) has 92 protons, 140 neutrons and 92 electrons

6 Copy and complete the table to show the subatomic particles in these elements.

Element	Protons	Neutrons	Electrons	Mass number
Li	**a**	4	**b**	7
Na	**c**	**d**	11	23
Kr	36	**e**	**f**	84
Pb	**g**	**h**	82	207

7 Chlorine comes in two forms $^{35}_{17}Cl$ and $^{37}_{17}Cl$.
a How do you know that they are both the same element?
b What is the difference between the two versions, or isotopes, of chlorine?
c Which element is X an isotope of? Explain your answer. $^{12}_{6}C$ $^{14}_{6}X$ $^{14}_{7}N$

8 **a** Write down the atomic number and mass number for the first 10 elements. (Use the Periodic Table on page 59.) Calculate the number of neutrons for each of the first 10 elements.
b Plot a graph of the number of neutrons (on the y-axis). Draw a line of best fit.
c What is the approximate relationship between the number of protons and the number of neutrons for the first 10 elements?
d Does this simple relationship hold for larger atoms such as uranium ($^{238}_{92}U$) and lead ($^{207}_{82}Pb$)? Explain your answer.

3 Atomic structure

1 Copy and complete these sentences. Use the words below to fill the gaps.

eight electrons energy shell

The .... in an atom are not free to move where they like. They can only occur in fixed electron .... positions (.......................... levels). The first shell can only take two electrons, while the second shell can take up to .... electrons.

2 Match these elements to their electron shell diagrams. Write each element's name, number and electron shell pattern in the following form:

calcium ($^{40}_{20}Ca$) 2,8,8,2

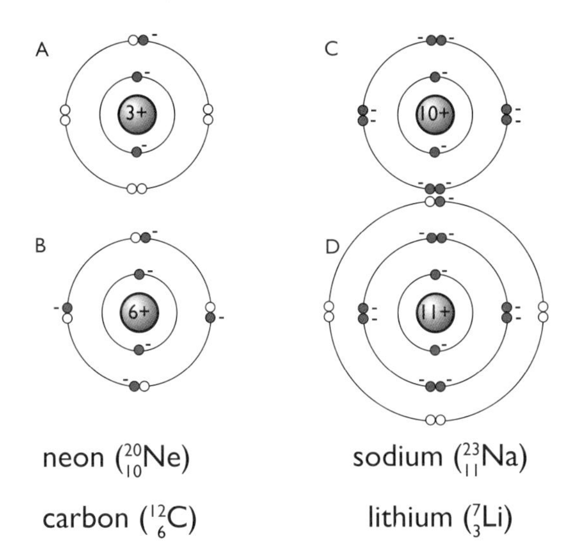

neon ($^{20}_{10}Ne$) sodium ($^{23}_{11}Na$)

carbon ($^{12}_{6}C$) lithium ($^{7}_{3}Li$)

3 Using diagrams like the ones in Q2, draw the electron positions for boron (5), oxygen (8), magnesium (12), argon (18) and potassium (19).

4 Look at the electron shell diagrams for atoms of the three unreactive gases helium ($_2He$), neon ($_{10}Ne$) and argon ($_{18}Ar$).

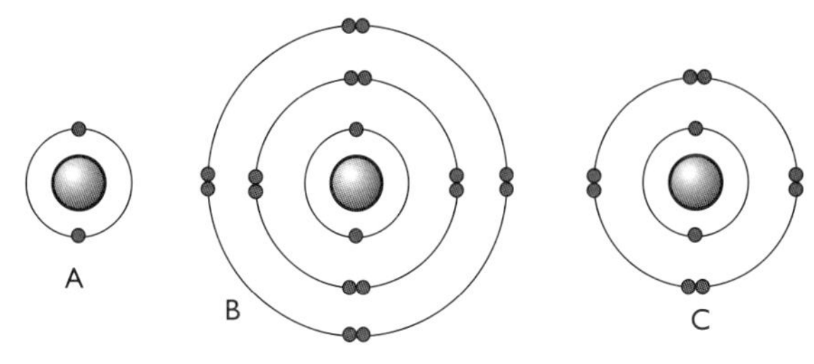

a Copy out the three diagrams and name them.

b What have the three atoms got in common?

c A full outer shell gives great stability. What chemical property does this give to these three gases?

d Draw an electron shell diagram for a sodium (atomic number = 11) ion, Na^+.

e Draw an electron shell diagram for a fluorine (atomic number = 9) ion, F^-.

f What do you notice about your diagrams for parts **d** and **e**? Which unreactive gas are they like?

g Suggest a reason why sodium and fluorine atoms form ions like this.

h What is the difference between the sodium and fluorine ions?

5 When chlorine is bubbled through sodium hydroxide it produces bleach, sodium chloride and water. The equation for the reaction is as follows:

$$2NaOH + Cl_2 \rightarrow NaClO + NaCl + H_2O$$

Steve claims that the mass of the products of this reaction will be greater than the mass of the reactants. He says that there are three products and only two reactants, so they must weigh more. Sami disagrees and thinks the mass of the reactants and products will be the same. Who is right, and why?

4 Solids, liquids and gases

1 Copy and complete these sentences. Use the words below to fill the gaps.
closer gases particles solids squashed

Everything is made up of tiny .... In .... and liquids, these particles are close together, so solids and liquids cannot be .... In ...., these particles are far apart. When you squash a gas, you push the particles .... together.

2 Copy and complete the table, writing **fixed** or **not fixed** in the missing slots.

	Shape	Volume
solid	**a**	fixed
liquid	**b**	fixed
gas	not fixed	**c**

3 **a** Copy the diagrams, showing how the particles are arranged in solids, liquids and gases. Label them correctly as solid, liquid or gas.

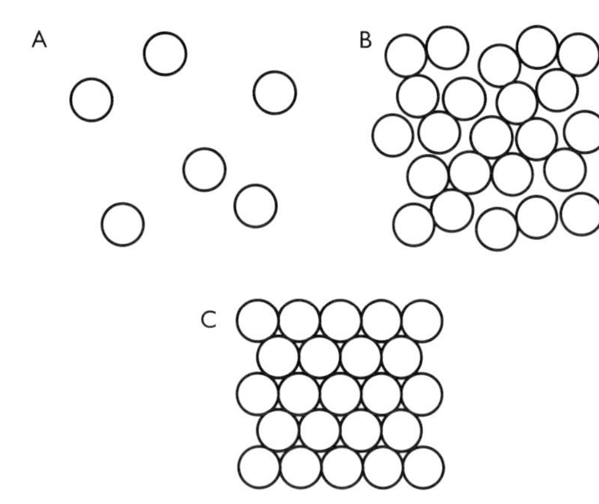

b Which is the most ordered arrangement?
c Which has the biggest gap between the particles?
d In which are the particles held firmly in place?
e In which two are the particles free to move?
f In which are the particles the most free to move?

4 In every substance, it is as if there is a battle going in between forces that hold the particles together and movement which tries to tear them apart. Copy and complete the following sentences, choosing the correct ending from below.
a In a solid …
b In a liquid …
c In a gas …

Choose endings from
- it is evenly balanced. The forces still keep the particles together, but they can slip and slide over one another.
- motion has won! The particles have broken free and are whizzing about at high speed.
- the forces are winning. The particles can only vibrate in their fixed places.

5 Students set up the apparatus shown below. Through the microscope the pollen grains appear to be moving about. This movement was seen for the first time in 1827 by the botanist Robert Brown.

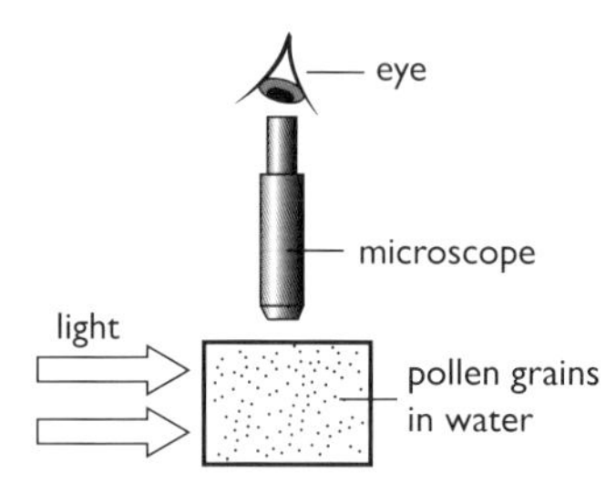

a What is this movement called?
b How can it be used to explain the movement of particles of matter?

5 Keep them moving

1 **a** Rearrange these sentences to explain how ice turns to water as it is heated. Copy the sentences out in the correct order.

- As ice is heated, the particles vibrate faster and faster.
- The ice melts.
- At 0 °C, the particles are vibrating fast enough to start snapping the force bonds that hold them together.
- In solid ice, the particles are vibrating about fixed positions.

b What is the melting point of ice?

c Sulphur melts at 113 °C. Do you think the forces between sulphur particles are stronger or weaker than those between water particles in ice? Explain your answer.

2 **a** Rearrange these sentences to explain how steam turns to water as it cools. Copy the sentences out in the correct order.

- At or below 100 °C, the particles stick together if they collide.
- The water particles in steam are far apart and moving very fast.
- The steam condenses.
- If the particles collide they simply bounce apart again.
- Clumps of particles stick together and collect, forming liquid water droplets.
- As the steam cools, the particles slow down.

b Alcohol vapour only condenses if the temperature drops below 77 °C. Do you think the forces between alcohol particles are stronger or weaker than those between water particles in steam? Explain your answer.

c Explain what happens to the particles in water if they are heated to 100 °C.

3 Dr Martin played a nasty trick on his class. He let off a stink bomb at the front without telling them. He timed how long it took for different pupils to notice.

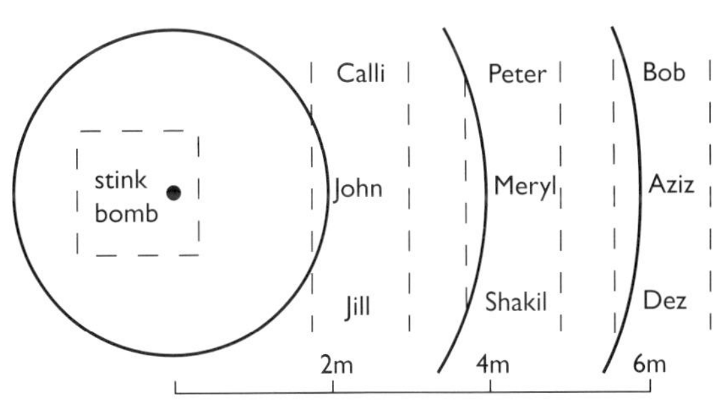

Here are his results:

John	(10 seconds)	Shakil	(23 seconds)
Calli	(16 seconds)	Aziz	(29 seconds)
Meryl	(20 seconds)	Bob	(33 seconds)

a How did the 'smell particles' reach the pupils?

b Explain briefly how this happens.

c Why did John notice first?

d Plot a graph of distance in metres against time in seconds for the smell of travel (estimate distance from the circles). Draw a line of best fit.

e Use this to estimate the 'time of arrival' of the smell to Jill, Peter and Dez.

f How fast did the smell diffuse through the class (in metres per second)? (speed = distance/time)

4 **a** Copy the diagram that shows how salt dissolves in water. Use the sentences below to label your diagram.

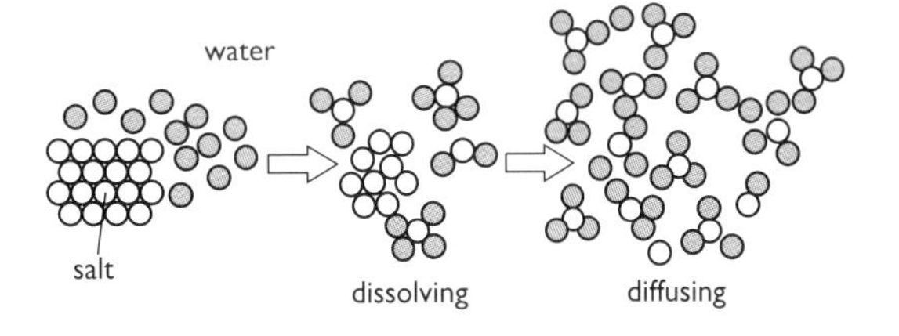

- The salt and water particles diffuse away, allowing more salt particles to be freed, and so on.
- Some of the outer salt particles are jostled free.
- Water particles surround the salt crystal.

b Is this diffusion faster or slower than in a gas? Explain your answer.

6 Covalent bonding

1 **a** What is a covalent bond?

b Using the data sheets on page 60, work out the electronic structure of the following non-metals. For each one, state how many electrons it would need to share in a covalent bond.

i oxygen (O) **ii** sulphur (S)
iii fluorine (F) **iv** carbon (C)

2 Represent the covalent bond in each of the following molecules in three different ways.

a hydrogen chloride
b oxygen
c chlorine
d methane

3 **a** Explain the characteristic melting and boiling points of covalent substances.

b Most covalent substances do not conduct electricity. Explain why.

4 Draw dot and cross diagrams to show the following.

a carbon + oxygen → carbon dioxide
b nitrogen + hydrogen → ammonia
c the structure of ethane (C_2H_6).

5 **a** Copy the diagram and use the idea of sharing to explain how chlorine atoms can join together to form a Cl_2 molecule.

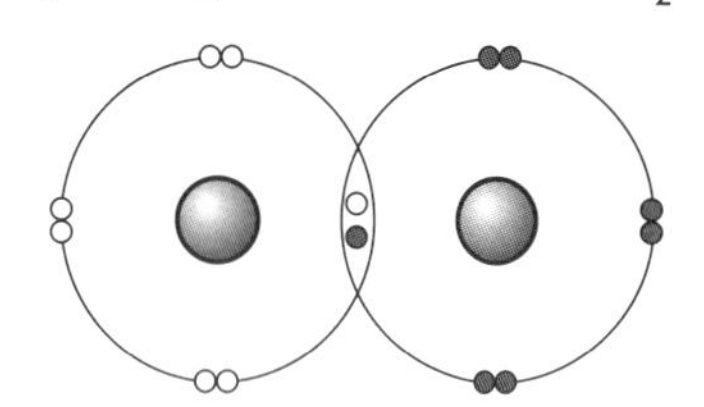

b What is this kind of bond called?

The diagram above part **b** shows just the outer electron shells.

The diagram below is a 'dot and cross' diagram that shows the outer shell electrons from one atom as 'dots' and from the other as 'crosses'.

c Draw a 'dot and cross' diagram to show how an F_2 fluorine molecule could form (atomic number = 9).

d Oxygen is O 2,6. How many electron pairs will oxygen need to share to form a covalent O_2 molecule?

e Draw a 'dot and cross' diagram for an O_2 molecule.

f Draw 'dot and cross' diagrams for

i water, H_2O (H = 1, O = 8)
ii ammonia, NH_3 (N = 7)
iii methane, CH_4 (C = 6)
iv carbon dioxide, CO_2

(There are only two electrons in the first shell.)

7 Ionic bonding

1 Copy and complete these sentences. Use the words below to fill the gaps.

carbon ions metals molecules

Compounds made from non-metallic elements, such as ...., hydrogen and oxygen, form uncharged particles called ....When .... form compounds with non-metals, they form ionic compounds which involve charged particles called ....

2 Copy and complete the sentence choosing the correct word from each pair. You should be able to make two correct sentences.

The charge on a **metallic/non-metallic** ion is **negative/positive**.

3 Different atoms can make different numbers of bonds. The diagrams show some simple compounds between metals and non-metals. The dots and crosses represent the number of electrons needed or available to transfer to produce stable structures.

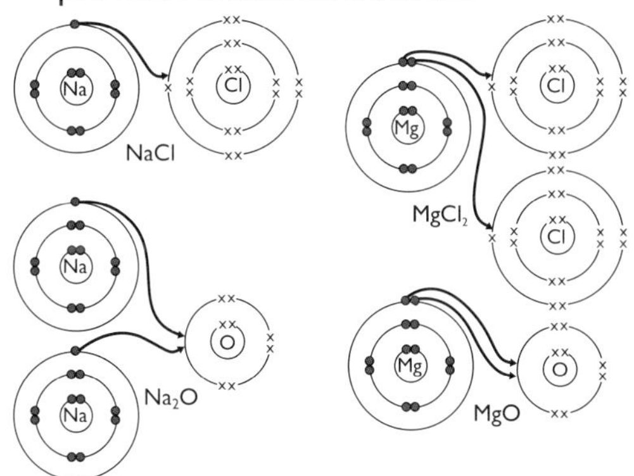

a Potassium (K) has one 'spare' electron, while calcium (Ca) has two. Use this idea to work out the formulae of

i potassium chloride

ii potassium oxide

iii calcium chloride

iv calcium oxide.

b Lithium chloride has the formula LiCl. How many spare electrons does lithium have in the outer shell?

c Lead chloride has the formula $PbCl_2$. How many electrons does lead lose here?

d Sodium bromide is NaBr. How many electrons have been transferred to bromine (Br) here?

Some metals can vary the number of bonds they make. For example, iron can have two, written as iron(II), or three, written as iron(III).

e What are the formulae of

i iron(II) chloride **ii** iron(III) chloride

iii copper(I) oxide **iv** copper(II) oxide?

4 **a** How do the melting and boiling points of covalent and ionic substances differ?

b Explain the characteristic melting and boiling points of ionic substances.

5 **a** The electron configuration of sodium can be written as Na 2,8,1. What part of this tells you that sodium is a metal?

b What must sodium do to form an ion?

c Chlorine can be written as Cl 2,8,7. What part of this tells you that chlorine is a non-metal?

d Rearrange these sentences to explain how sodium and chlorine atoms combine to form sodium chloride. Copy them out in the correct order.

- An ionic bond is formed, making sodium chloride.
- The oppositely charged ions are attracted to each other.
- The chlorine atom accepts the electron.
- It becomes a negative ion.
- The sodium atom donates its 'loose' electron becoming a positive ion.

6 The electron configuration of calcium can be written as Ca 2,8,8,2.

a How do you know that calcium is a metal?

b What must it do to form an ion?

c What will the charge on the ion be?

d How many chlorine ions could be made using the outer electrons from one calcium atom?

e From your answer to part **d**, what must the formula of calcium chloride be?

f Draw electron shell diagrams to show how calcium chloride could be formed.

8 Metallic bonding and intermolecular forces

1 Copy and complete these sentences. Use the words below to fill the gaps.

covalent giant high low molecules strong weak

Although .... bonds are very strong, the forces between covalent molecules are .... Because of this, substances with small ...., such as methane or ammonia, have very .... melting and boiling points. Some covalent materials, such as diamond or silicon dioxide, form .... structures. Because every bond in these materials is a .... covalent bond, they are hard solids with .... melting and boiling points.

2 The diagram shows how the particles are arranged in solid iron and solid lead. The forces holding the iron particles together are stronger than the forces holding the lead particles together.

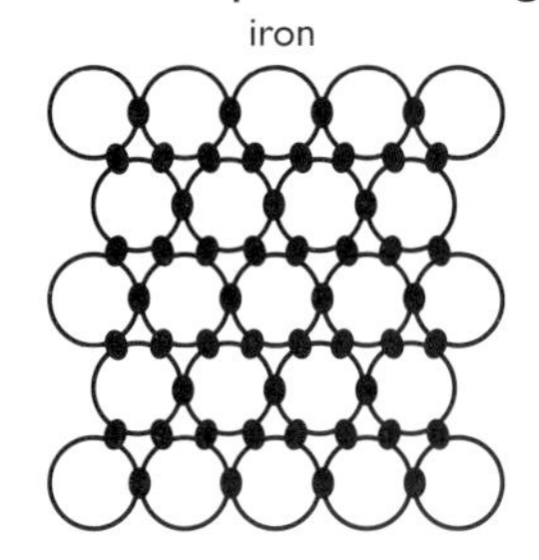

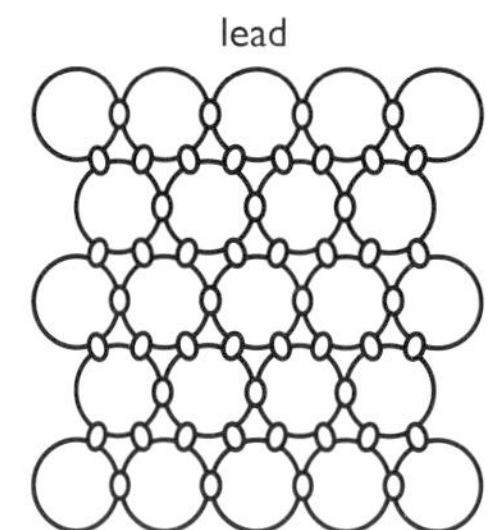

o ● represent the forces holding the particles together

Copy and complete the following sentences, choosing the correct ending from on the right.

a Solids like iron and lead keep their shape because ...

b Iron is harder and stronger than lead because ...

c Lead is heavier for its size than iron because ...

Choose endings from

- the forces between the particles in iron are stronger than lead.
- lead particles weigh more than iron particles.
- the particles are held in place by forces.

3 **a** Ionic substances are often crystalline. Covalent substances sometimes form crystals as well. Give three examples of covalent substances which form simple molecular crystals.

b How do they form these crystal structures?

4 The diagram shows the arrangement of water molecules in three different states.

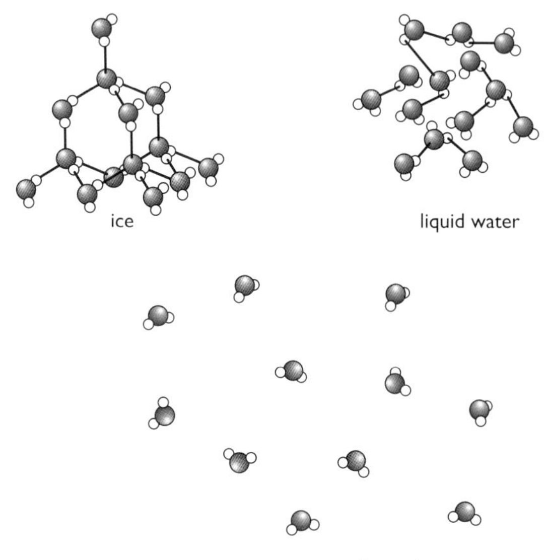

a Explain why ice (solid water) floats on top of water (liquid water) in terms of the distance between water molecules.

b Explain what happens to the intermolecular forces between the molecules when water boils, and when the water vapour cools down below 100 °C.

9 Structures

1 **a** Which properties would you expect in simple covalent molecular structures?
b Explain how the structure of simple molecular compounds explains these properties.

2 **a** What are allotropes?
b Diamond and graphite are allotropes of which element?
c Give the two main allotropes of phosphorus.

3 **a** Draw and describe the shapes of the following molecules.
i carbon dioxide **ii** water
iii ammonia **iv** methane
b What affects the shape of covalent molecules like these?

4 **a** Which type of structure is formed when carbon dioxide, water, ammonia and methane are cooled down?
b What is unusual about the element iodine?
c Describe the difference between what happens when you heat solid water (ice) and when iodine is heated.

5 **a** Explain the main differences in the properties of diamond and graphite and explain why they are so different.
b How do the structures of diamond and graphite affect how they are used?

10 Structures and bonding

1 Explain why substances consisting of giant structures usually have high melting and boiling points.

2 Look at the diagram of diamond and graphite.

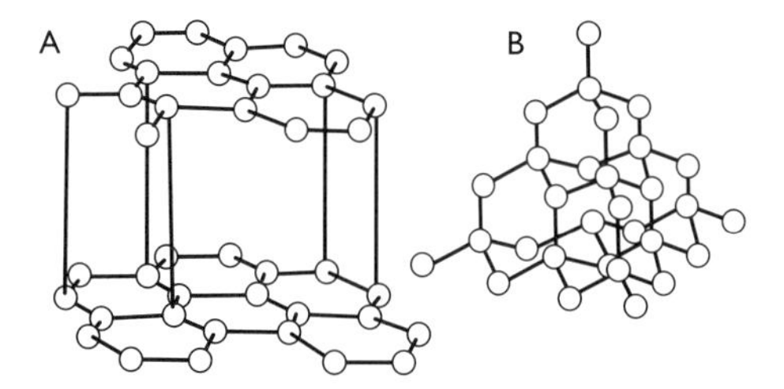

a Copy and complete: A is ...., B is ....

b What is the same about these two giant structures, and what is different?

c Explain why both diamond and graphite have very high melting points.

d Explain why diamond is very hard, but graphite is very soft.

3 Ionic compounds, such as sodium chloride, form giant structures.

a Which force holds the ions together in sodium chloride?

b What properties do these bonds give ionic solids?

c Ionic substances are made of charged particles. Why do they not conduct electricity when they are solid?

d What are the two ways that the ions can be freed up, so that the ionic material can conduct?

e We write the formula of the compound sodium chloride as NaCl. Does such a unit actually exist? Explain your answer.

4 Copy the diagram of a metallic giant structure.

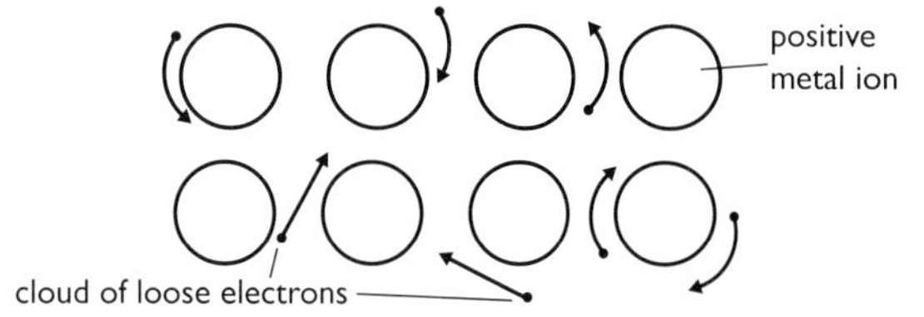

a How does this explain the hardness and high melting point of most metals?

b How does it explain the fact that solid metals conduct electricity?

c How does it explain the fact that metals can be easily shaped without breaking?

5 The table headings are four different types of structures which you might find in common materials. Copy the table and complete it to answer these questions.

Structure	Ionic	Molecular	Metallic	Giant molecular
melting and boiling points				
conduction when solid				
conduction when liquid				
example				

a Under each heading, write whether you would expect to see high or low melting and boiling points.

b Write whether you would expect a solid to conduct electricity.

c Write whether you would expect a liquid to conduct electricity.

d Give one example of a substance with each type of structure.

6 **a** Copy and complete this table.

Substance	Melting point (°C)	Boiling point (°C)	Electrical conductivity at room temperature	Electrical conductivity when liquid	Structure
quartz (silicon dioxide)	1610	2230			
rubidium fluoride	795	1410			
manganese	1244	1962			

b Explain the difference in electrical conductivity between the three substances shown in the table.

11 Separating mixtures

❶ Copy and complete these sentences. Use the words below to fill the gaps.
atoms compound different element

All substances are made from .... There are over 90 .... kinds of atom. A substance made from one kind of atom only is called an ....A substance made from two or more different types of atoms joined together is called a ....

❷ This apparatus can be used to separate a mixture of sand and salt.

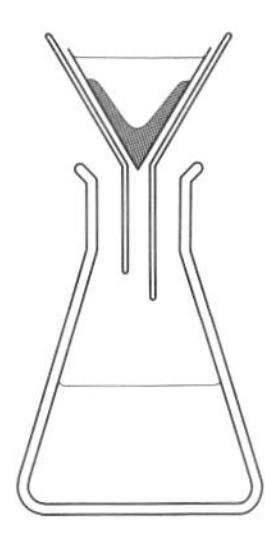

a What is this process called?
b Write a set of instructions to show exactly how this should be done, including a labelled diagram.
c To complete the process, you will need to remove the salt from the water. Explain carefully, using a diagram, exactly how this would be done and what the process is called.

❸ **a** Which technique would you use to separate two substances in a mixture when they both dissolve in water?
b Explain carefully how this technique works.

❹ This apparatus is used to separate a mixture of two substances, such as water and ink dye.

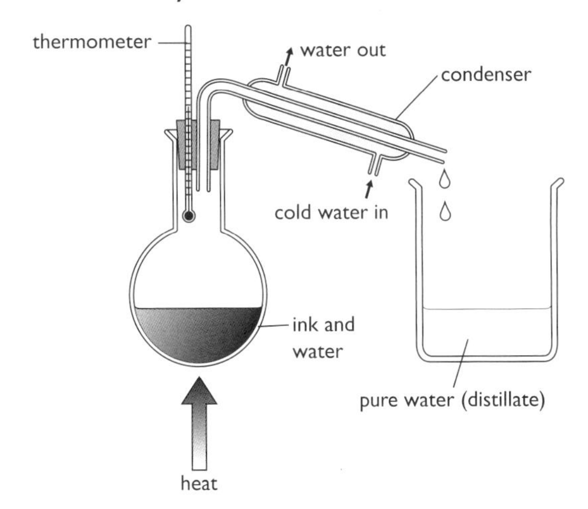

a What is the name of this process?
b What is the purpose of the thermometer?
c What is the purpose of the condenser and why does it have cold water running through it?

❺ Children's sweets are often very brightly coloured. You want to show that the blue colouring in a particular brand of sweets is actually made up of several different dyes.
a Describe how you might do this.
b Explain how this works.

1 A compound's name can tell you which elements have combined. So sulphur dioxide is a compound made from sulphur and oxygen atoms. The name dioxide also tells you that there are two oxygen atoms present. This can be shown in a diagram, or by the chemical formula, SO_2.

Copy and complete the table of some simple compounds. Match the name to its diagram and formula below. (Alternative names have been given where the common names do not follow the simple rules.)

Compound	Diagram	Formula
hydrogen chloride	**a**	**b**
carbon dioxide	**c**	**d**
carbon monoxide	**e**	**f**
ammonia (nitrogen trihydride)	**g**	**h**
water (dihydrogen oxide)	**i**	**j**

Choose from the following diagrams and formulae.

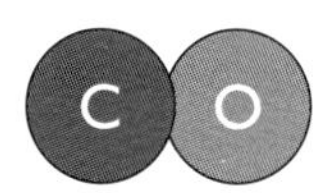

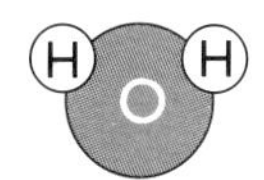

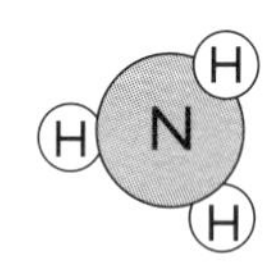

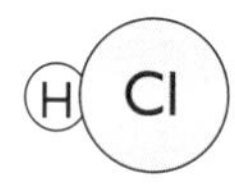

CO_2 HCl H_2O NH_3 CO

2 For each of these compounds, write down which elements make up the compound and how many of each type there are.

a methane, CH_4 (natural gas)
b octane, C_8H_{18} (in petrol)
c glucose, $C_6H_{12}O_6$
d sulphuric acid, H_2SO_4
e nitric acid, HNO_3

3 Match the correct formula to each compound and then copy them out.

a	calcium carbonate	$PbBr_2$
b	sodium chloride	Al_2O_3
c	aluminium oxide	NaCl
d	lead bromide	$CaCO_3$

4 Carbon and oxygen (the reactants) combine to give carbon dioxide (the product).

a Draw a table with two columns labelled 'Reactants' and 'Products'. For each of the following reactions, put the substances in the correct columns.
  A If you burn hydrogen in oxygen you get water.
  B When magnesium reacts with sulphuric acid you get magnesium sulphate and hydrogen.
  C Rust is a form of iron oxide which you get when iron reacts with oxygen in the air.
  D If you heat copper carbonate you get copper oxide and carbon dioxide.

b Write out each reaction as a word equation, in the form carbon + oxygen → carbon dioxide.

5 **a** The symbol (aq) means in solution in water (from the Latin for water, which is *aqua*). Suggest what (s), (l) and (g) mean.

b Look at the following balanced chemical equations. For each, write out the reaction as a word equation.

i $2Mg(s) + O_2(g) \rightarrow 2MgO(s)$
ii $Na_2O(s) + 2HCl(aq) \rightarrow 2NaCl(aq) + H_2O(l)$
iii $CuO(s) + H_2SO_4(aq) \rightarrow CuSO_4(aq) + H_2O(l)$
iv $2Al(s) + Fe_2O_3(s) \rightarrow 2Fe(l) + Al_2O_3(s)$
v $Mg(s) + 2HCl(aq) \rightarrow MgCl_2 + H_2(g)$
vi $2Na(s) + 2H_2O(l) \rightarrow 2NaOH(aq) + H_2(g)$

13 Chemistry by numbers

1 Look at the diagram.

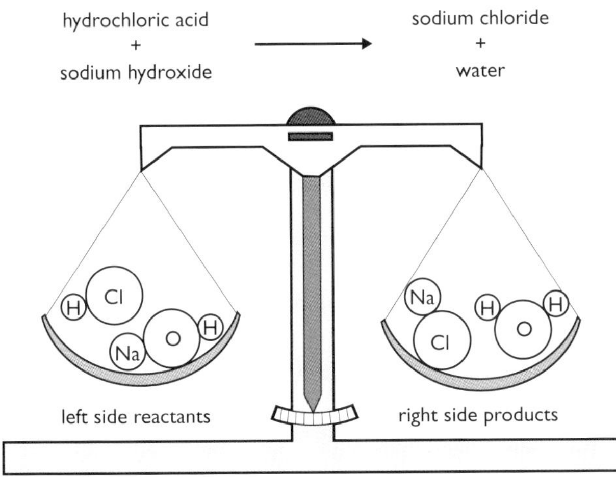

a balanced equation

a Write this out as a balanced chemical equation.

b Copy and complete this sentence, choosing the correct bold word in each case. The correct sentence states a fundamental rule of chemistry.

The **colour/mass/volume** of the reactants is equal to the **colour/mass/volume** of the products.

c Draw a diagram similar to the one above for the reaction when copper carbonate breaks down on heating.

$CuCO_3(s) \rightarrow CuO(s) + CO_2(g)$

d Write this reaction out as a word equation.

e 12.4 g of copper carbonate was heated. When it was reweighed, its mass was only 8 g. How does this fit the rule in part **b**?

f What mass of carbon dioxide must have been formed?

2 Copy and complete these equations, adding the missing mass values (in g).

a magnesium + oxygen → magnesium oxide
(2.4 g) (1.6 g) (… g)

b silver bromide → silver + bromine
(1.88 g) (1.08 g) (… g)

c sodium + chlorine → sodium chloride
(… g) (7.1 g) (11.7 g)

3 For each pair below, write out equation **i** as a word equation and write a balanced equation for reaction **ii**.

a **i** $2Na + Br_2 \rightarrow 2NaBr$
ii sodium reacts with chlorine (Cl) to give sodium chloride

b **i** $2Mg + O_2 \rightarrow 2MgO$
ii calcium (Ca) burns in oxygen to give calcium oxide

c **i** $4Na + O_2 \rightarrow 2Na_2O$
ii potassium (K) burns in oxygen to give potassium oxide

4 Here are some unbalanced equations.
i Write out each reaction as a word equation.
ii Balance each chemical equation.

a $CaO + HCl \rightarrow CaCl_2 + H_2O$
b $K + H_2O \rightarrow KOH + H_2$
c $CaCO_3 + HCl \rightarrow CaCl_2 + H_2O + CO_2$
d $Mg + HCl \rightarrow MgCl_2 + H_2$

5 Atoms gain or lose electrons when they form ions. For example, an aluminium atom loses three electrons to form an Al^{3+} ion. This can be written as $Al \rightarrow Al^{3+} + 3e^-$ (electrons). Write similar ionic equations for the following changes.

a a sodium (Na) atom becoming an Na^+ ion
b a copper (Cu) atom becoming a Cu^{2+} ion
c a chlorine *atom* becoming a Cl^- ion
d a chlorine *molecule* (Cl_2) becoming *two* Cl^- ions

14 Calculating chemicals

Use the data tables on page 60 to help you answer some of these questions.

❶ Copy and complete these sentences. Use the words below to fill in the gaps.

compound elements react relative atomic mass

The atoms of different .... all have different masses. So that we know how the masses of different atoms compare with each other, we use their .... (A_r). Then we can work out the relative formula mass (M_r) of a ...., which is very useful when we are measuring out substances to .... together.

❷ **a** What is meant by the relative atomic mass of an element?

b What is meant by the term weighted average?

❸ The relative formula mass is found by adding together the relative atomic masses of all the atoms in a molecule. Calculate the relative formula mass of each of the following substances.

For example

water, H_2O

$2 \times A_r$ of hydrogen = $2 \times 1 = 2$

$1 \times A_r$ of oxygen = $1 \times 16 = 16$

So the M_r of water is $2 + 16 = 18$

a ammonia, NH_3

b magnesium chloride, $MgCl_2$

c copper sulphate, $CuSO_4$

d ethanol, C_2H_5OH

❹ **a** Work out the relative formula mass of these substances.

CO	Fe	CaH_2	PH_3
N_2H_4	H_2O_2	CH_3OH	N_2
H_2CO_3	NaF	CaO	MgF_2

b Put the substances in pairs, where each member of the pair has the same relative formula mass.

15 More chemical calculations

Use the data tables on page 60 to help you answer these questions.

1 **a** Ammonium nitrate, NH_4NO_3, and ammonium phosphate, $(NH_4)_2HPO_4$ are both artificial fertilisers. Calculate the percentage of nitrogen in each fertiliser.

b Farmer Smith found the yield of his cereal crop was lower than usual last year. The plants showed symptoms of lacking nitrogen. If he used the same amount of each of these two fertilisers, which would be most effective for improving the crop yield. Explain your answer.

2 Pure silicon is extracted from silicon dioxide.

a If the silicon is extracted from 240 g of silicon dioxide, what mass of silicon would you expect? (Show all your working.)

b In a commercial setting, if 360 tonnes of silicon dioxide is processed, what mass of silicon would you expect?

3 **a** What is a mole of any substance?

b Write down the simple formula you can use to help you calculate the number of moles of a substance.

c You are asked to measure out 3 moles of NaCl. What mass of NaCl do you need?

d You are given 320 g of copper sulphate ($CuSO_4$). How many moles of copper sulphate are there?

4 One mole of an element contains 6.02×10^{23} atoms and has the same mass as its relative atomic weight in grams. For example, one mole of oxygen contains 6.02×10^{23} atoms and has a mass of 16 g, whilst one mole of hydrogen contains 6.02×10^{23} atoms and has a mass of 1 g. Calculate the number of atoms in the following:

a 23 g of sodium

b 355 g of chlorine

c 6.4 g of sulphur

d 60 g of magnesium

e 0.06 g of carbon

5 **a** What is the empirical formula of a compound?

b What is the molecular formula of a compound?

c In a reaction 2.4 g of magnesium were reacted with oxygen in the air, 4.0 g of magnesium oxide was produced. What is the empirical formula of the product?

16 Calculations using moles

Use the data tables on page 60 to help you answer these questions.

1 Patronite is an ore which contains vanadium combined with sulphur as vanadium sulphide. If 3.58 g of patronite contain 1.02 g of vanadium, what is the empirical formula of patronite?

2 Magnetite is an ore of iron which contains 72.4% by mass of iron, the rest being oxygen. What is the empirical formula of magnetite?

3 We know that the mass of a chemical = number of moles present × mass of 1 mole. Put another way

$$\text{number of moles of a chemical} = \frac{\text{mass of chemical}}{\text{mass of 1 mole of the chemical}}$$

Use this equation to calculate the number of moles in the following compounds.

a 62 g of Na_2O **b** 22 g of CO_2
c 5.8 g of KF **d** 30 g of $MgSO_4$
e 6.75 g of $CuCl_2$

4 Powdered aluminium and iron(III) oxide react vigorously according to the following equation:

$Al + Fe_2O_3 \rightarrow Al_2O_3 + Fe$

a Balance this equation.
b In one reaction, 5 moles of aluminium were mixed with 2 moles of iron(III) oxide. How many grams of iron metal could be produced using these quantities of reactants?

5 Phosphoric acid, H_3PO_4, is used to make phosphate fertilisers. Phosphoric acid can be made by boiling phosphorus oxide, P_4O_{10}, with water.

$P_4O_{10} + H_2O \rightarrow 12\ H_3PO_4$

a Balance this equation.
b If 14.2 g of P_4O_{10} were reacted with water how much phosphoric acid was formed **i** in moles **ii** in grams?

6 Sodium carbonate, Na_2CO_3, is an important chemical used in many industrial processes. It is made from salt, using a method called the Solvay process. This has two steps:

(1) $H_2O + NaCl + NH_3 + CO_2 \rightarrow NH_4Cl + NaHCO_3$

(2) $2NaHCO_3 \rightarrow Na_2CO_3 + CO_2 + H_2O$

a Rewrite the first equation to show the quantities of all the chemicals which would have to react to give $2NaHCO_3$ as one of the products.
b How many moles of sodium carbonate could be made from 100 moles of salt?
c How many moles of carbon dioxide are required when 4 moles of salt react to form sodium carbonate?
d How many grams of salt would be needed to make 31.8 g of sodium carbonate using the Solvay process?

7 **a** What is meant by the term molar volume?
b What is the difference between the molar volume of a gas at room temperature and pressure and a gas at standard temperature and pressure?
c At what temperature and pressure are standard conditions measured?
d Sulphur dioxide (SO_2) is a gas. You are given 0.5 moles of the gas at room temperature and pressure. What volume will it take up and what mass of gas have you got?
e Use your answer from part **d** to calculate the density of sulphur dioxide.

8 Calculate the volume of hydrogen evolved at stp when excess hydrochloric acid is added to 4 g of zinc.

$Zn(s) + 2HCl(aq) \rightarrow ZnCl_2(aq) + H_2(g)$

17 Metals, air and water

❶ Copy and complete each sentence using the correct ending from below.

a Copper metal …
b Gold metal …
c Sodium metal …
d When a metal reacts with air it …
e If a metal reacts with water it …

Choose endings from

- reacts very quickly with air.
- tarnishes slowly in air.
- 'steals' the oxygen from water, leaving hydrogen.
- does not tarnish in air.
- combines with the oxygen in the air to form an oxide.

❷ Lead and copper have both been used for water pipes (although lead is no longer used as it is poisonous). Why are lead and copper suitable for carrying water, but iron and magnesium are not?

❸ Copper is usually used to conduct electricity in wires and cables. However, the contacts in certain switches and in computer equipment use gold. Explain this.

❹ Potassium, calcium and zinc all react with oxygen and with water. The products of these reactions are sodium oxide, calcium oxide and zinc oxide for the reactions with oxygen, and sodium hydroxide, calcium hydroxide and zinc oxide for the reactions with water. Write word equations for all these reactions.

❺ **a** Explain what is meant by the reactivity series.
b Explain how a reactivity series is drawn up by observing how different elements react with air, water and acids.

18 Displacement reactions

❶ Copy and complete these sentences. Use the words below to fill the gaps. Each word may be used more than once.

copper displacement reactivity zinc

When a more reactive metal is dipped in a solution containing a less reactive metal, a .... reaction takes place. An example of this type of reaction is when a piece of .... is dipped in .... sulphate solution, where the .... metal displaces the .... from solution. Studying these reactions enables us to draw up a .... series.

❷ Aluminium reacts with iron oxide in a reaction called the thermite reaction. In this reaction, the aluminium displaces the iron. The reaction releases a great deal of energy, leaving the iron liquid.

a Write a word equation for this reaction.

b The reaction is used to weld the ends of rails together when a railway track is laid. What makes it suitable for this purpose?

❸ Zinc will displace copper from a solution of copper sulphate, and magnesium will displace zinc from a solution of zinc sulphate.

a Write word equations for these reactions.

b Write down a reactivity series for these three metals, putting the most reactive metal first.

❹ Some Martian school pupils carried out displacement reactions for the metals scrittiby, splerbity, snibitty, stobbity, slibbity and blib and their snerbide solutions. They collected the results shown in the table, although they did not finish their investigation. If displacement took place they showed it by a ✓, if it did not, they put a ✗.

a Copy and complete the table as far as you can.

Metal	Solution					
	scrittiby snerbide	splerbity snerbide	snibitty snerbide	stobbity snerbide	slibitty snerbide	blib snerbide
scrittiby	–	✗				✓
splerbity		–	✓			
snibitty			–		✓	
stobbity	✓	✓	✓	–	✓	✓
slibbity	✗	✗			–	
blib		✗	✗		✓	–

b As far as you can, draw up a reactivity series for the metals.

c What further investigation(s) would you need to do to produce a complete reactivity series?

❺ **a** The chemical formulae for magnesium sulphate, copper sulphate and zinc sulphate are $MgSO_4$, $CuSO_4$ and $ZnSO_4$ respectively. Write balanced chemical equations for the displacement reactions between magnesium, zinc and copper and their solutions. Use the state symbols (aq) to indicate a solution ($MgSO_4(aq)$ for example) and (s) to indicate a solid ($Zn(s)$ for example).

b Why are displacement reactions so important in the development of a reactivity series?

❻ **a** What does the term oxidation mean?

b What does the term reduction mean?

c What is meant by a redox reaction?

❼ **a** Look at the reactivity series on the data tables on page 60. Which metals do you think can be extracted from their ores using carbon?

b Which type of reactions are these and why?

c Why isn't carbon used to extract silver, gold and platinum?

❽ **a** Which of the following reactions are reduction reactions?

i $CuO + H_2 \rightarrow Cu + H_2O$

ii $Fe_2O_3 + 3CO \rightarrow 2Fe + 3CO_2$

iii $NaOH + HCl \rightarrow NaCl + H_2O$

iv $SnO_2 + 2Mg \rightarrow Sn + 2MgO$

b Write down the ionic equations for equations **i** and **iv.**

19 Acids and alkalis

1 Copy and complete each sentence using the correct ending from below.

a Substances which change colour according to whether they are in acid, neutral or alkaline solutions are …

b When a substance dissolves in water it forms a solution which may be …

c The pH scale is …

d When non-metal oxides dissolve in water their solutions are …

e When metal oxides dissolve in water their solutions are …

Choose endings from

- acidic, neutral or alkaline.
- acidic, with a pH less than 7.
- called indicators.
- alkaline, with a pH greater than 7.
- used to show how acidic or alkaline a solution is.

2 Choose the correct symbol which is used to indicate a substance that is corrosive to living tissues.

Copy the symbol and label your drawing 'The symbols used to show that a substance attacks and destroys living tissues, including eyes and skin. A substance like this is called corrosive.'

3 Copy the scale representing the pH scale.

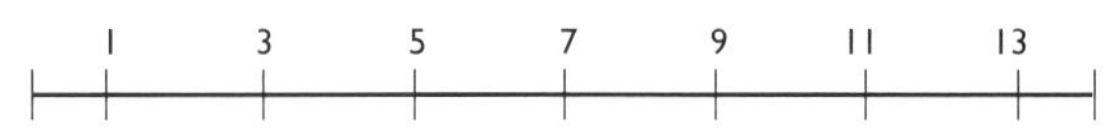

Mark on your scale the approximate pH of

a vinegar (a weak acid)

b pure water

c caustic soda (a strong alkali)

d hydrochloric acid (a strong acid)

e toothpaste (a weak alkali)

f What is the difference between a strong acid and a weak acid? Give examples of both types which are used in the science laboratory.

4 Not all indicators change colour at the same pH. The table shows some indicators, the colour change which occurs as the pH increases (the solution becomes more alkaline) and the approximate pH at which this happens.

Indicator	Colour change as pH increases	Approximate pH at which change occurs
methyl orange	red → orange	4
litmus	red → blue	5
bromothymol blue	yellow → blue	7
phenolphthalein	clear → pink	9

a Using your scale from **Q3**, add coloured arrows to show where each indicator changes colour.

b Which indicator could be used to distinguish a solution which is weakly acidic from a solution which is strongly acidic?

c Which indicator could be used to distinguish a solution which is weakly alkaline from a solution which is strongly alkaline?

d Which indicator could be used to distinguish a solution which is weakly alkaline from a solution which is weakly acidic?

20 Metals and acids

1 Copy and complete these sentences. Use the words below to fill in the gaps.

bubbles of gas dissolves
hydrogen reactive smaller

When a metal reacts with an acid .... are seen. These are .... As a piece of metal reacts, it gets .... as it .... in the acid. The more .... the metal, the faster it reacts.

2 Copy and complete each sentence using the correct ending from below.

a All acids are …
b When an acid is dissolved in water it forms …
c When an alkali dissolves in water it forms …
d When an acidic solution reacts with an alkaline solution …
e An acid reacting with an alkali is called …

Choose endings from

- hydrogen ions, H^+.
- hydroxide ions, OH^-.
- substances which contain hydrogen.
- neutralisation.
- H^+ and O^- ions combine to form water.

3 When metals react with an acid, a gas is given off. Describe carefully the test you would do on a test tube of this gas to show what it is.

4 Give a definition for

a an acid **b** a base
c an alkali **d** a salt.

5 Three metals A, B and C were each put into some strong acid. The amount of gas produced in each tube was measured, and the graphs plotted.

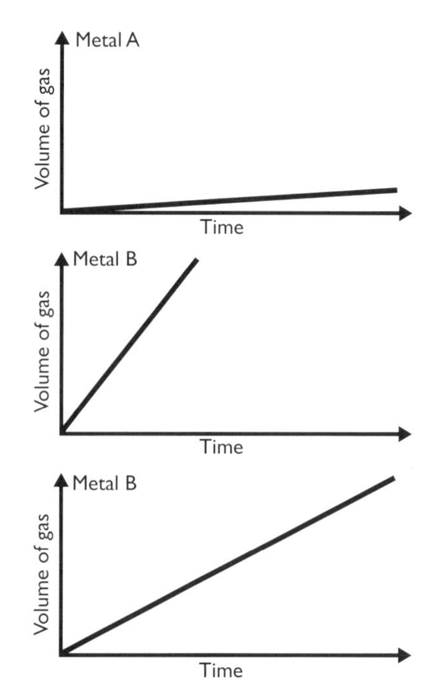

a Which metal produced hydrogen quickest?
b Which metal produced hydrogen slowest?
c Put the three metals in order of reactivity, most reactive first.
d How would you make sure that this investigation was a fair test?

6 a Using words and diagrams, describe how you would investigate the reaction of the metals copper, magnesium and zinc with hydrochloric acid (a strong acid) with this equipment.

b Give word and balanced chemical equations for each of these reactions.

21 Neutralisation

1 Copy and complete the following sentences. Use the words below to fill in the gaps.

indigestion metal neutralisation salt

When acids and bases react, the reaction is called .... This can be summarised as
acid + base → .... + water.

Bases are often .... compounds such as oxides, hydroxides, hydrogen carbonates and carbonates. Bases such as magnesium hydroxide are used in medicines to cure ....

2 '**B**icarbonate for **b**ees, **v**inegar for **v**asps' (wasps!) is one way to remember how to treat bee and wasp stings. What does this tell you about the pH of bee and wasp stings?

3 Chalk (calcium carbonate) is sometimes dumped into lakes affected by acid rain.

a Burning fossil fuels produces sulphur dioxide. Why can this produce acid rain?

b Why can it be helpful to add chalk to lakes affected by acid rain?

c Why not use sodium hydroxide instead of chalk?

4 Write word equations to describe the products of the following reactions.

a zinc oxide + hydrochloric acid

b magnesium oxide + sulphuric acid

c copper carbonate + nitric acid

d sodium carbonate + ethanoic acid

5 Give a general equation for each of the following reactions. For each reaction give an example.

a a metal oxide and an acid

b a metal hydroxide and an acid

6 Give word and balanced equations for the reactions between an acid and the hydroxide of an alkali metal that would result in the following salts.

a potassium sulphate

b lithium chloride

c sodium nitrate

22 Salt formation

1 Describe carefully how you would produce a pure sample of copper sulphate crystals using this equipment.

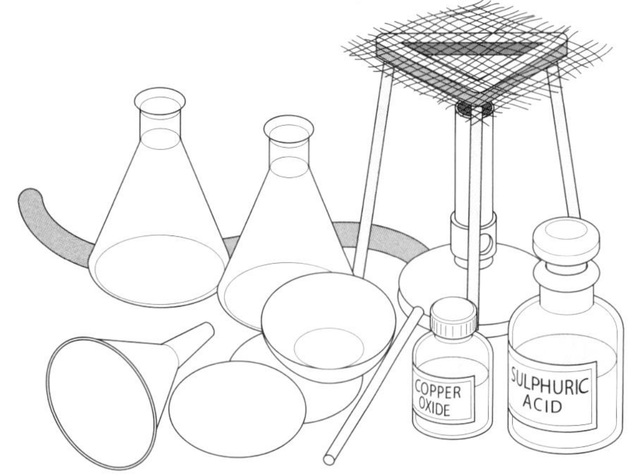

2 Acid salts are salts that contain some hydrogen ions. One example of an acid salt is sodium hydrogen sulphate, $NaHSO_4$, a white powder found in some lavatory cleaners. Sodium hydrogen sulphate is especially good for cleaning lavatories and basins in hard water areas, where limescale (calcium and magnesium carbonates) builds up.

- **a** Explain why sodium hydrogen sulphate is useful in hard water areas.
- **b** Write an equation to explain what happens when sodium hydrogen sulphate is poured onto limescale. (Hint: treat sodium hydrogen sulphate as a mixture of sodium sulphate (which is not involved in the reaction) and sulphuric acid.)
- **c** Why is water needed before sodium hydrogen sulphate can act as a cleaner in this situation?

3 The diagram shows three pairs of beakers. Copy each pair of beakers, and draw a third beaker in which the contents of the first two beakers are mixed. Draw lines between the ions that react.

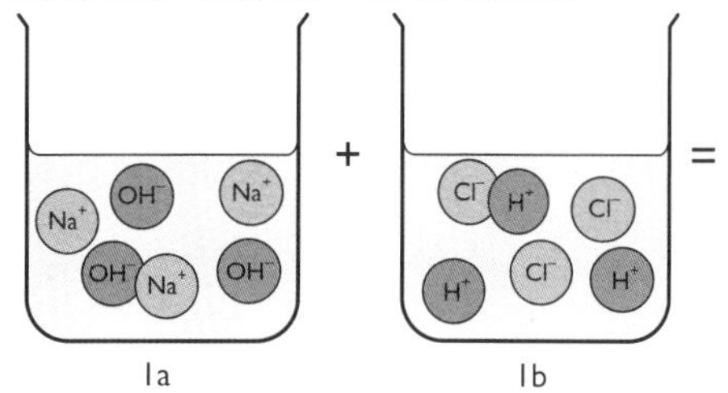

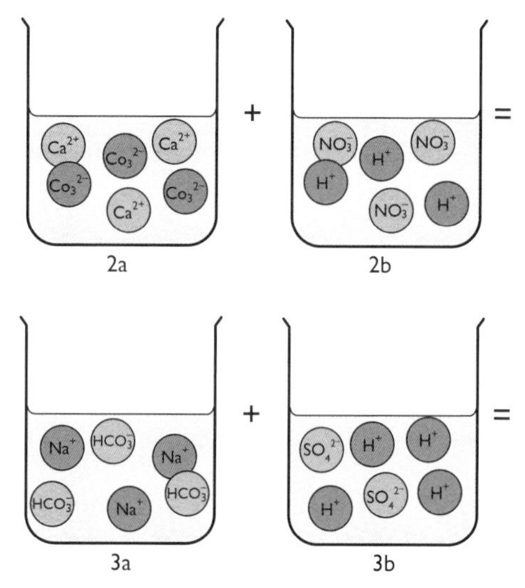

4
- **a** Describe an experimental procedure for making the transition metal salt of your choice, using a metal oxide or hydroxide and an acid.
- **b** Give one advantage of using insoluble compounds in a reaction like this.
- **c** Give one disadvantage of using insoluble compounds.

5
- **a** Which salts are all soluble?
- **b** Which salts are least soluble?
- **c** Give balanced chemical equations showing the state of the reacting substances for these reactions.
  - **i** iron(III) oxide + hydrochloric acid
  - **ii** copper hydroxide + nitric acid

6 Salts are made using one of four main methods.

- **a** What are these methods?
- **b** For each of these methods, choose a suitable salt and describe in detail how it would be made.

7 Write ionic equations for the reactions in **Q3**.

8 When you make copper sulphate you get copper sulphate solution. If you heat this solution to dryness, you get a white powder. If you leave the water to evaporate away slowly, you get blue crystals. These two substances are both copper sulphate. Explain the difference.

23 Titration calculations

Use the data tables on page 60 to help you answer some of these questions.

1 Explain carefully how you would carry out a titration.

2 **a** What is the concentration of a solution when 40 g of sodium hydroxide are dissolved in water to make 1 dm^3 of sodium hydroxide solution?
b What is the concentration of the sodium hydroxide solution if the 40 g of NaOH was added to water to make 500 cm^3 of solution?

3 A student carried out a titration using 25.0 cm^3 sodium hydroxide solution of unknown concentration, which was placed in a conical flask. The sodium hydroxide was exactly neutralised by 20.0 cm^3 of 0.50 mol dm^{-3} hydrochloric acid added from a burette. What was the concentration of the sodium hydroxide solution?

4 In a titration, 15.0 cm^3 of hydrochloric acid reacted exactly with 10.0 cm^3 of sodium hydroxide solution. The concentration of the acid was 0.10 mol dm^{-3}.
a Write an equation for this reaction.
b Calculate the number of moles of hydrochloric acid in the acid solution added to the sodium hydroxide solution.
c Write down the number of moles of sodium hydroxide in the sodium hydroxide solution.
d Calculate the concentration of the sodium hydroxide solution.

5 Vinegar contains ethanoic acid. A 20.0 cm^3 sample of vinegar was titrated against 0.50 M sodium hydroxide solution. Exactly 25.0 cm^3 of the sodium hydroxide solution was needed to neutralise the ethanoic acid in the vinegar.
a Write an equation for this reaction.
b Calculate the number of moles of NaOH in the sodium hydroxide solution added to the vinegar.
c Write down the number of moles of ethanoic acid in the vinegar.
d Calculate the concentration of the ethanoic acid in the vinegar.

6 A sample of water taken from a lake was found to contain sulphuric acid. A student carried out a titration to find the concentration of the sulphuric acid in the sample. 25.0 cm^3 of the sulphuric acid was neutralised exactly by 34.0 cm^3 of a potassium hydroxide solution of concentration 0.2 mol dm^{-3}. The equation for the reaction is

$$2KOH(aq) + H_2SO_4(aq) \rightarrow K_2SO_4(aq) + 2H_2O(l)$$

a Describe the experimental procedure which would be used for this titration.
b Calculate the number of moles of potassium hydroxide used.
c Calculate the concentration of the sulphuric acid in mol dm^{-3}.

7 2.5 g of washing soda (sodium carbonate) was dissolved in water in a conical flask. Through titration it was found that exactly 17.5 cm^3 of hydrochloric acid with a concentration of 1.0 mol dm^{-3}.
a Suggest an indicator which could have been used in the titration.
b Write a balanced equation for the reaction between hydrochloric acid and sodium carbonate.
c Calculate the number of moles of hydrochloric acid which reacted with the sodium carbonate in the flask.
d How many moles of sodium carbonate was this equivalent to?
e What mass of sodium carbonate was this?
f What mass of the water softener in the conical flask was water?
g How many moles of water was this?
h The formula of sodium carbonate can be written as $Na_2CO_3.xH_2O$.
Use your answers to parts **d** and **g** to calculate a value for *x*.

24 Rates of reaction

1 Copy the table and complete it to show which of the chemical reactions are fast and some which are slow.

rusting of metal **oil forming**
fireworks going off **an explosion**
silver tarnishing **coal burning**

Slow reactions	Fast reactions

2 The diagram shows a conical flask containing hydrochloric acid on a top pan balance. Thomas drops a strip of magnesium ribbon into the acid. It fizzes, and eventually dissolves. Thomas notices that the reading on the balance has decreased.

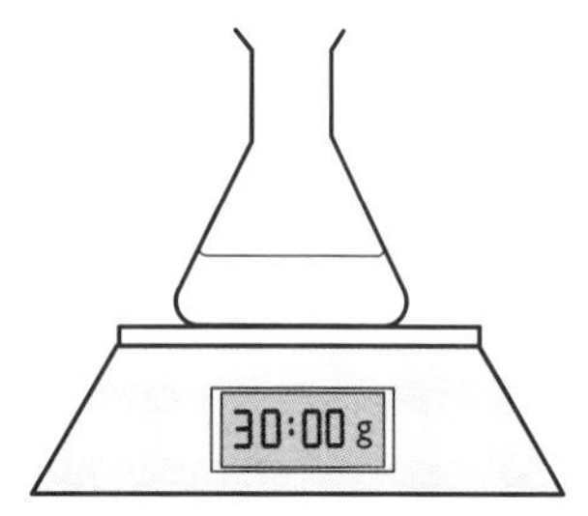

a Why did the reading on the balance decrease?

b As well as the reading on the balance, what else would Thomas need to record in order to measure the rate of this chemical reaction?

c When magnesium reacts with hydrochloric acid it is a vigorous, fizzing, spitting reaction. The decrease in the reading on the balance was very small. Why is it a good idea to include a cotton wool bung in the neck of the flask when carrying out this experiment?

3 The graph shows the rate at which the mass of a flask containing calcium carbonate and hydrochloric acid decreases as the reaction progresses.

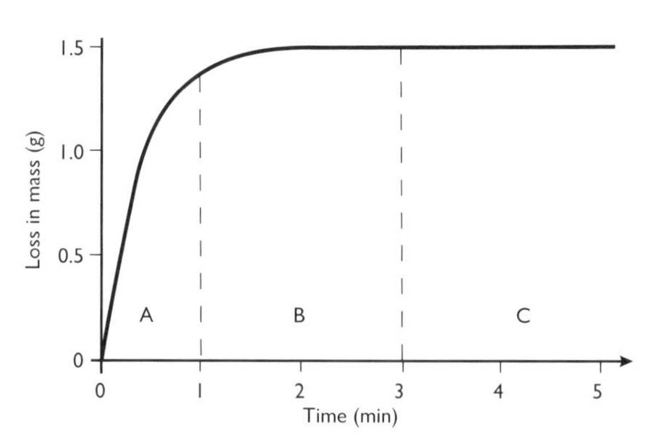

a Suggest why the reaction mixture loses mass as the reaction goes on.

b Explain the shape of the graph in the areas marked A, B, and C.

4 Winston, Mona and Sherena carried out an experiment in which they added a piece of magnesium ribbon to hydrochloric acid in a flask on a top pan balance. They timed the loss of mass as the reaction took place and recorded their results in the table above.

Time (s)	Mass of flask + contents (g)	Total loss of mass (g)
0	170.00	0.00
10	169.96	0.04
20	169.92	0.08
30	169.88	0.12
40	169.85	0.15
50	169.83	0.17
60	169.82	0.18
70	169.81	0.19
80	169.80	0.20
100	169.80	0.20

a Plot a line graph of these results.

b What was the total mass of hydrogen produced in this reaction?

c How long did it take for this hydrogen to be formed?

d Use your answers to parts **b** and **c** to help you work out the average rate of the reaction in g/min.

25 What affects chemical reactions?

❶ Copy and complete these sentences. Use the words below to fill the gaps.

area concentration double pressure smaller

The rate of a chemical reaction involving a solid can be increased by making the surface .... of the solid bigger. This can be done by using .... lumps of the solid. If a reaction involves a solution, increasing the .... of the solution will increase the rate of reaction. When gases react together, the rate of reaction can be increased by increasing the .... Temperature also affects the rate of a chemical reaction – a 10 °C increase in temperature will roughly .... the rate of a reaction.

❷ Explain the following.

a Small sticks catch fire much more quickly than a big log.

b Food keeps longer in a freezer than in a refrigerator.

c A splint burns steadily in air, but flares up when put into a gas jar of oxygen.

d Advertisers claim that a soluble painkiller acts more quickly than one taken as a tablet.

❸ Great care must be taken in sawmills where wood is sawn into planks, since explosions happen easily if there is a spark or a flame. Use the following statements to help you explain this. Write your answer as complete sentences.

- Tiny particles of sawdust spread out through the air.
- An explosion is really just a very rapid burning reaction.
- Sawing wood produces tiny particles of sawdust.
- Wood needs to be heated before it will start to burn.
- The smaller the pieces of solid in a reaction, the faster the reaction.
- Wood burns in air.

❹ Sodium thiosulphate reacts with hydrochloric acid, producing a precipitate of sulphur which makes the reacting mixture go cloudy. The following table shows how long it takes for a cross marked on a piece of paper under the flask to become invisible. The same concentration of reactants was used in each case.

Temperature (°C)	Time for cross to become invisible (s)
15	200
25	100
35	50
45	25
55	12.5

a Plot a graph of these results.

b What does your graph show about the effect of temperature on the time it takes for the cross to disappear?

c What does this tell us about the effect of temperature on the rate of the reaction?

d Why does temperature have this effect on the rate of a reaction?

5 Explain carefully how the following factors affect the rate of reactions. Use a kinetic model of chemical reactions in your explanations.

a particle size

b concentration

c temperature

26 Catalysts and enzymes

1 Copy and complete each sentence using the correct ending from below.

- **a** An enzyme is …
- **b** The activation energy is …
- **c** A catalyst is …
- **d** Chemical reactions can only happen when …
- **e** Concentration, surface area and temperature can all affect …

Choose endings from

- the minimum amount of energy particles must have to react.
- the rate of a chemical reaction.
- reacting particles collide with one another.
- a substance that can speed up the rate of a chemical reaction.
- a biological catalyst.

2 Copy and complete these sentences, using the words below to fill in the gaps.

affected enzymes faster lower metals rate

Catalysts increase the .... of a chemical reaction without altering anything else. They can be used to make a reaction go .... or they can be used to make a reaction happen at a .... temperature. They are not .... themselves and can be used time after time. Catalysts are often .... In living cells special biological catalysts called .... control all the chemical reactions which go on.

3 Look at the diagram that shows some metal catalysts used in a chemical reaction. Explain why the pieces of catalyst have holes in them, using your knowledge about things that affect the rate of chemical reactions.

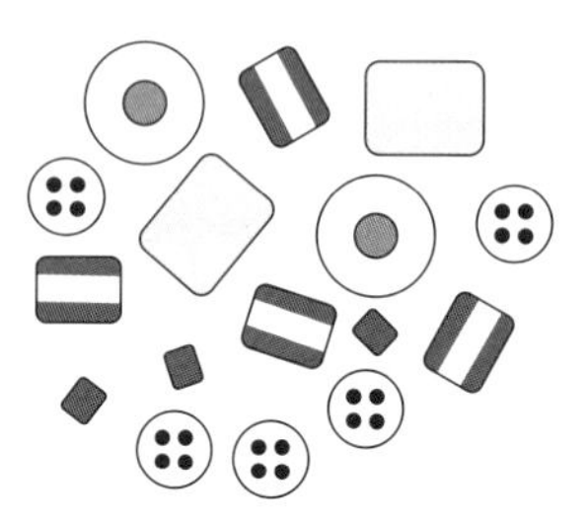

4 The graph shows the rate of a chemical reaction which can be catalysed by three different enzymes. The enzymes come from three different living things – one is a bacterium which lives in hot springs, another is a mammal, and the third is a fish which lives in Arctic waters.

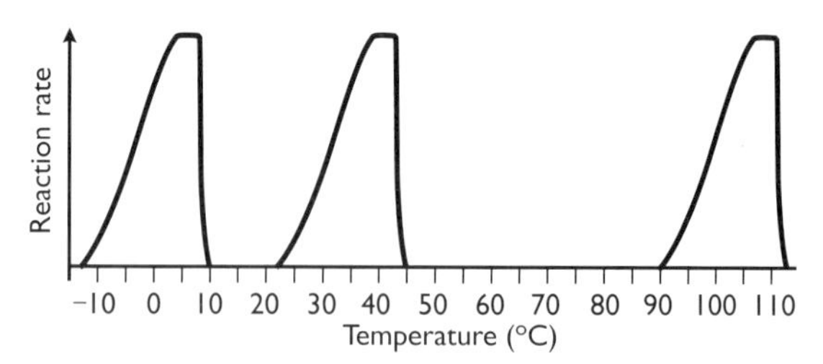

- **a** Copy the graph and label it clearly to show which enzyme comes from which living thing.
- **b** Each line showing enzyme activity has a very similar shape. Explain what the shape tells us about how enzymes work.

27 Activation energy

1 Explain why

a a lower activation energy means a chemical reaction goes faster

b a higher activation energy means a chemical reaction goes slower.

2 a Copy and complete the energy diagrams, using these labels. You may use each label more than once.

reactants energy products
activation energy (without catalyst)
activation energy (with catalyst)
course of reaction

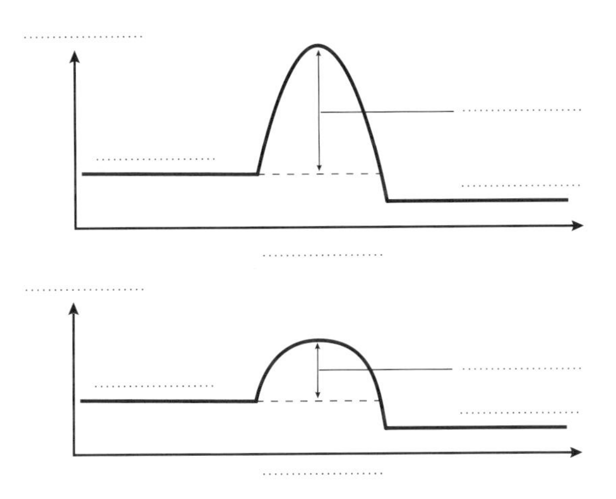

b Use your labelled diagrams to explain how a catalyst speeds up a chemical reaction.

3 Catalysts speed up the rate of chemical reactions without being changed themselves. They are often found as very small beads or pellets with holes in them. Use your knowledge of chemical reactions and what affects the rate of those reactions to answer these questions.

a How do catalysts increase the rate of chemical reactions?

b Why are catalysts usually found as tiny beads or pellets with holes in them?

28 Energy changes in reactions

❶ Copy and complete each sentence using the correct ending from below.

- **a** Chemists describe chemical reactions …
- **b** Chemical reactions need energy to be supplied …
- **c** Energy can be supplied in the form of heat or electricity…

Choose endings from

- to make chemical reactions happen.
- using equations.
- before they will happen.

❷ In an exothermic reaction energy is 'given out', or transferred, to the surroundings. In an endothermic reaction energy is 'taken in' from the surroundings.

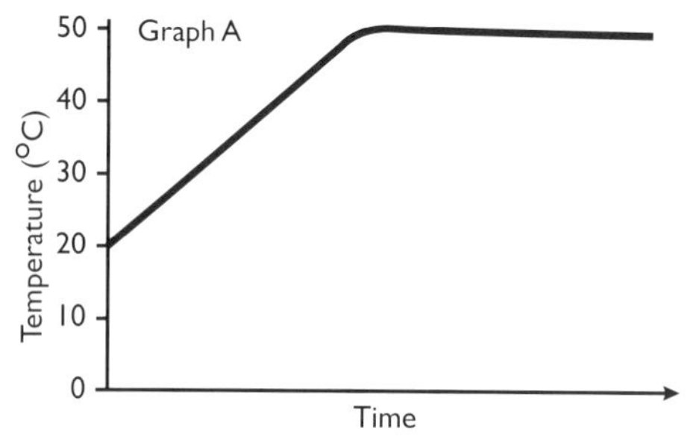

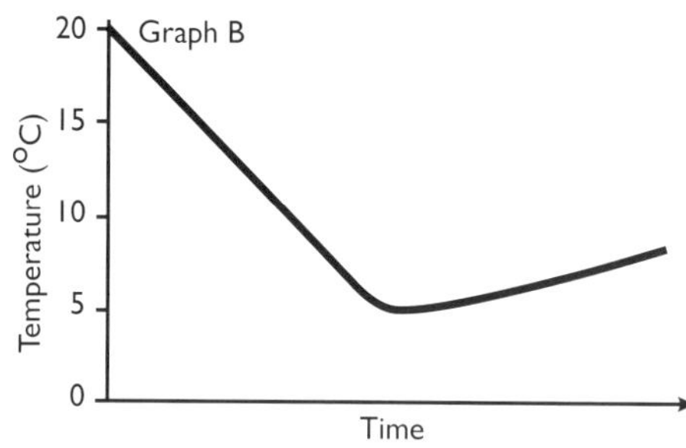

- **a** Does graph A show an endothermic or an exothermic reaction? How do you know?
- **b** Does graph B show an endothermic or an exothermic reaction? How do you know?
- **c** Which of the following chemical changes are exothermic and which are endothermic?
  - **i** a candle burning
  - **ii** ammonium chloride dissolving, producing a decrease in temperature
  - **iii** dynamite exploding
  - **iv** respiration **v** photosynthesis

❸ Sherbet is a mixture of sugar and a chemical called sodium hydrogen carbonate. A pupil has an idea that the feeling of 'fizziness' when you put sherbet in your mouth is caused by an exothermic reaction. Design an experiment that the pupil could do to test this idea.

❹ Dynamite is an explosive which contains atoms of the elements carbon, hydrogen, nitrogen and oxygen. When it explodes, it produces nitrogen gas, carbon dioxide gas, water (as steam) and oxygen.

- **a** Write a word equation for the explosion of dynamite.

The person in the diagram is pushing down the plunger of a detonator. This sends an electric current along two wires, which sets off a small explosive charge. This detonates the dynamite that blows up the building.

- **b** Why must the dynamite be detonated before it will explode?
- **c** Suggest another way of detonating dynamite.
- **d** Why is exploding dynamite so destructive?

❺ When an acidic solution and an alkaline solution neutralise each other, a good deal of energy is released.

- **a** What effect does this energy have on the two solutions being mixed?
- **b** Design a simple experiment to demonstrate to some younger pupils that neutralisation reactions are exothermic (release energy). Use words and diagrams to explain your ideas.

29 Calculating energy changes

Use the data table of bond energies on page 60 to help you answer some of these questions.

1 **a** In any chemical reaction energy must be supplied in order to make a chemical reaction happen. What is this energy needed for?

b During a chemical reaction energy is released. What is happening as this energy is released?

c What is happening to the balance of these energy changes in

i an exothermic reaction

ii an endothermic reaction?

2 When William and Tanya heat chemical A it decomposes to form chemical B. They then react chemical B with chemical C to form chemical D. The table below gives some information about the energy changes in these chemical reactions.

Chemical reaction	Energy required for bond breaking (kJ/mol)	Energy released in bond making (kJ/mol)	Net energy transfer for reaction
A → B	350	500	
B + C → D	150	75	

a Copy and complete the table.

b What is the overall net energy transfer when chemical A produces chemical D in these two reactions?

3 **a** Copy and complete the table to show whether the chemical reactions are exothermic or endothermic.

Chemical reaction	ΔH (kJ/mol)	Exothermic or endothermic
ethanol + oxygen → carbon dioxide + water	−1367	
oxygen → ozone	+143	
nitrogen + hydrogen → hydrazine (rocket fuel)	+51	
magnesium + chlorine → magnesium chloride	−641	

b The reaction of oxygen to ozone happens in the upper atmosphere and needs sunlight. Why?

c Here is an energy level diagram for the reaction between ethanol and oxygen.

Draw an energy diagram for each of the three remaining reactions.

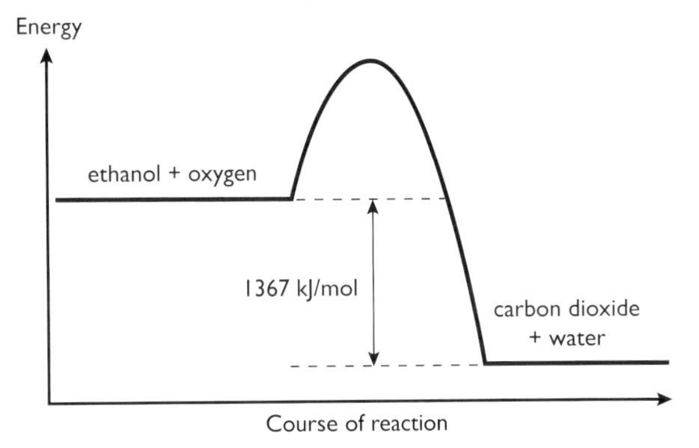

4 Sketch energy level diagrams for the following reactions.

a $6CO_2 + 6H_2O \rightarrow C_6H_{12}O_6 + 6O_2$ $\Delta H = +2880$ kJ mol^{-1}

b $C + O_2 \rightarrow CO_2$ $\Delta H = -394$ kJ mol^{-1}

c $H_2 + I_2 \rightarrow 2HI$ $\Delta H = +53$ kJ mol^{-1}

d $C_5H_{12} + 8O_2 \rightarrow 5CO_2 + 6H_2O$ $\Delta H = -3500$ kJ mol^{-1}

5 When hydrogen and chlorine react together they form hydrogen chloride.

a What are the energy changes for this reaction?

b State whether the reaction is endothermic or exothermic.

6 **a** What is meant by the term 'bond energy'?

b What is the difference between the bond energy for making and breaking bonds, and how do we indicate this difference?

7 Write balanced equations and then work out the net energy transfer for the following reactions. For each one state whether it is endothermic or exothermic.

a hydrogen + bromine → hydrogen bromide

b carbon + hydrogen → methane

c carbon monoxide + oxygen → carbon dioxide

d nitrogen + hydrogen → ammonia

30 Industrial chemistry

1 Copy and complete each sentence using the correct ending from below. Then arrange the sentences in the correct order to describe the production of ammonia.

- **a** The main ingredient of any nitrate fertiliser is …
- **b** Ammonia …
- **c** The raw ingredients of the Haber process are …
- **d** The nitrogen comes from the air …
- **e** Nitrogen is unreactive so the Haber process needs …
- **f** The process also uses a moderately high …

Choose endings from

- an iron catalyst.
- temperature and pressure.
- and the hydrogen comes from methane.
- is made by the Haber process.
- ammonia.
- nitrogen and hydrogen.

2 Copy the diagram of the Haber process. Use the labels below in place of A–F on your diagram.

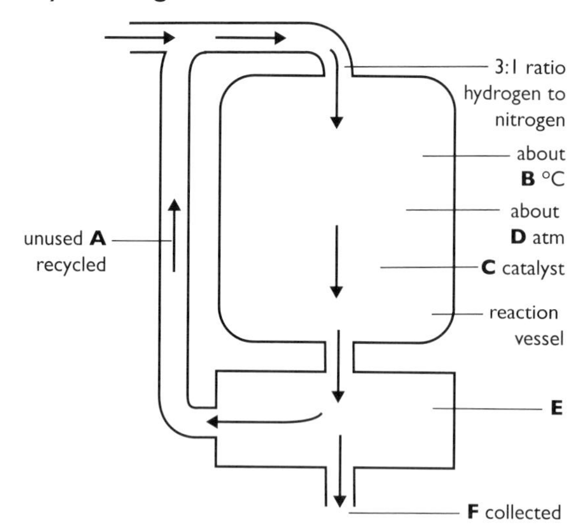

200 450 condenser hydrogen and nitrogen iron liquid ammonia

3 **a** What is the formula of sulphuric acid?
b Why is sulphuric acid such an important industrial product? Produce a flow chart summarising the different stages in the production of sulphuric acid.

4 At about 27 °C and a high pressure the yield of sulphuric acid in the Contact process would be almost 100%. Explain why these are not the operating conditions usually chosen when making sulphuric acid on an industrial scale.

5 The graph shows how temperature and pressure in the Haber process affects the yield of ammonia.

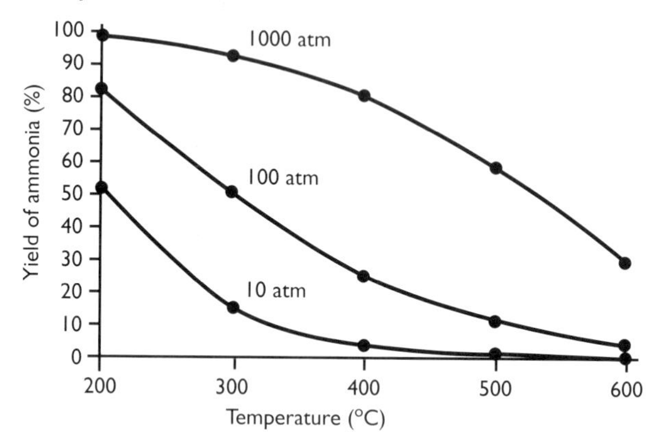

- **a** What is the yield of ammonia at 300 °C and at
  - **i** 10 atm pressure
  - **ii** 100 atm pressure
  - **iii** 1000 atm pressure?
- **b** Which conditions seem to give the best yield of ammonia?
- **c** Why are these not the normal operating conditions for the Haber process in industry?
- **d** What is the % yield of ammonia likely to be at the normal operating conditions of 450 °C and 200 atm and why are these conditions used?

31 The best conditions

1 Chemical plants have to make a profit, so the reactions need to run as quickly and efficiently as possible. Complete this table to give four factors which affect the rates of reactions along with an explanation of how the effect works.

Factor affecting rate of reaction	How it works

2 **a** Write a balanced chemical equation to describe the equilibrium that occurs when nitrogen and hydrogen react to produce ammonia.

b How many moles of gas are there on the left-hand side of the equilibrium?

c How many moles of gas are there on the right-hand side of the equilibrium?

d Does an increase in pressure increase or decrease the yield of products on the right-hand side of the equation?

e The usual operating pressure for the commercial production of ammonia is about 200 atm. Why this has been chosen rather than a higher or lower pressure?

3 Bloggs and Son are setting up a business to make chemical Z. Chemical Z is made when chemical X decomposes in the following reaction.

$X \leftrightharpoons Y + Z$

a What does the sign $\leftrightharpoons$ tell you about the reaction?

b If the reaction is allowed to take place in a closed vessel, equilibrium is reached. What does this mean?

c Mr Bloggs wants to make sure that he gets lots of chemical Z, and does not want the reaction to go into reverse and make chemical X again. What can he do to make sure that he gets as much Z as possible?

4 The diagram top right shows some apparatus that can be used to decompose ammonia into unreactive nitrogen and inflammable hydrogen.

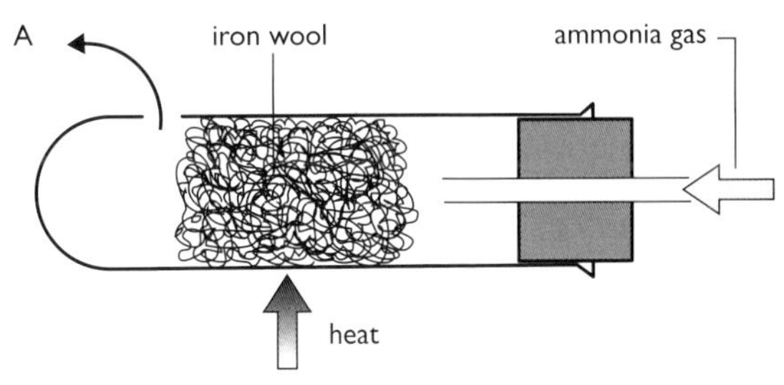

a What is the purpose of the iron wool?

b Most of the ammonia decomposes. Explain why.

c Suggest one simple chemical test to detect undecomposed ammonia in the stream of gases at A.

d How could hydrogen be removed from the mixture of gases leaving A?

5 What is the main difference between a reversible reaction and an ordinary reaction? Compare the reaction between iodine monochloride and chlorine with the reaction between hydrogen and chlorine to help with your explanation.

6 In a reversible reaction, if the reaction in one direction is exothermic, the reaction in the opposite direction will be endothermic. Using ideas of energy and chemical bonds, explain why the amount of energy transferred is always exactly the same in each direction.

7 **a** What is meant by a closed system?

b In a closed system reversible reactions reach a dynamic equilibrium. What does this mean?

8 Henri Le Chatelier said 'If the conditions of a reversible reaction are changed, the position of the equilibrium will shift to oppose that change.' Explain the effect of the following in terms of this statement.

a Temperature on the products of an exothermic reaction.

b Increased pressure on the products of the Haber process.

c Temperature on the products of an endothermic reaction.

32 Electrolysis

1 Copy and complete these sentences. Use the words below to fill in the gaps.
conductors current electrolysis electrolytes

Electrical energy is carried by an electric .... Metals are good .... of electricity. Electricity can also flow through solutions called .... Unlike metals, these are chemically changed when they conduct electricity. This process is called ....

2 a Copy and complete the diagram to show a solution being electrolysed. Use the labels below on your diagram.

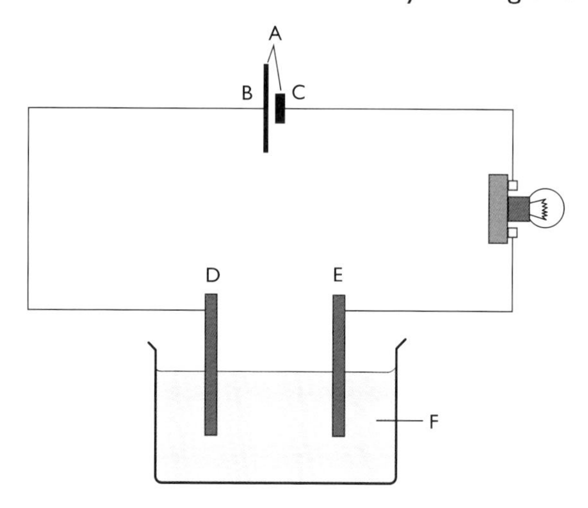

+ – battery anode (positive) cathode (negative) solution

b If this solution was water (H_2O), what would be produced
i at the anode ii at the cathode?

3 Match the name of the atom with the symbol for its ion.

bromine	O^{2-}	calcium	Pb^{2+}
Ca^{2+}	oxygen	hydrogen	Cu^{2+}
Cl^-	copper	Br^-	chlorine
Zn^{2+}	lead	zinc	H^+

4 Copy and complete the table to show what is formed when the solutions are electrolysed.

Solution	Cathode	Anode
copper chloride		
zinc bromide		
hydrochloric acid		

5 Draw up a table to show which of these ions would move towards the anode and which towards the cathode during electrolysis. (Use the data table on page 60 to help you.)
sodium iodide zinc iron oxide aluminium chloride fluoride silver

6 Copper can be purified by electrolytic refining, in which copper is removed from an impure anode, and is deposited on a pure copper cathode. The half-equation for the reaction at the anode is
$Cu(s) \rightarrow Cu^{2+}(aq) + 2e^-$

a Write down the half-equation for the reaction at the cathode.

b How do the electrons get from the anode to the cathode?

c One mole of electrons travels from the anode to the cathode. What mass of copper is deposited on the cathode? (The relative atomic mass of copper is 64.)

33 More electrolysis

1 Three important industrial chemicals can be made by electrolysing brine (sodium chloride solution). Copy the diagram of the electrolysis of brine.

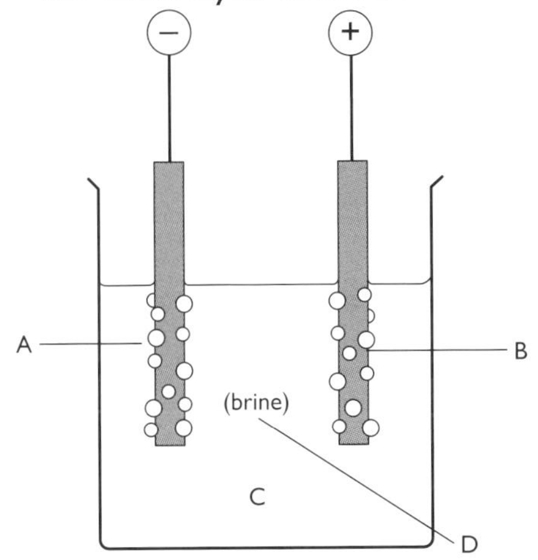

a Use these labels in place of A–D on your diagram.
- Positive sodium and hydrogen ions are attracted to the negative electrode – hydrogen gas forms.
- Negative chlorine ions are attracted to the positive electrode – chlorine gas is formed.
- Sodium chloride ionises when it dissolves in water.
- Sodium hydroxide is left in solution.

b Give the half equations to show what happens at the two electrodes during the electrolysis of brine.

c What provides the energy needed to tear sodium chloride apart in this way?

d Draw up a table to show the main uses of hydrogen, chlorine and sodium hydroxide.

2 a Explain how the apparatus shown top right can be used to demonstrate that ions move during electrolysis.

b Suggest another way in which you can demonstrate the movement of ions to the electrodes in electrolysis. Draw a diagram to help your explanation.

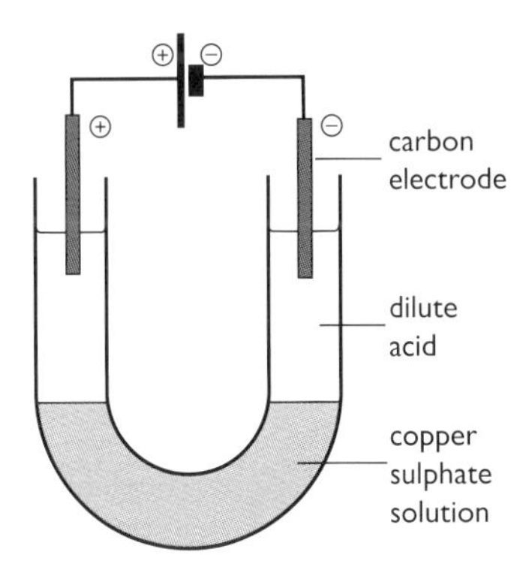

3 Once tin has been extracted from its ore it is still contaminated by trace elements and needs to be purified further before it can be used to plate steel cans to make 'tin' cans. Using your knowledge of the purification of copper, explain with a diagram how the tin might be purified.

4 a Draw a simple diagram of the electrolysis of molten lead bromide, showing what happens at the anode and the cathode.

b Give the half equations for the reactions at the anode and the cathode.

c Why are the reactions at the anode and the cathode known as oxidation and reduction reactions respectively?

5 The diagram shows the electrolysis of copper chloride.

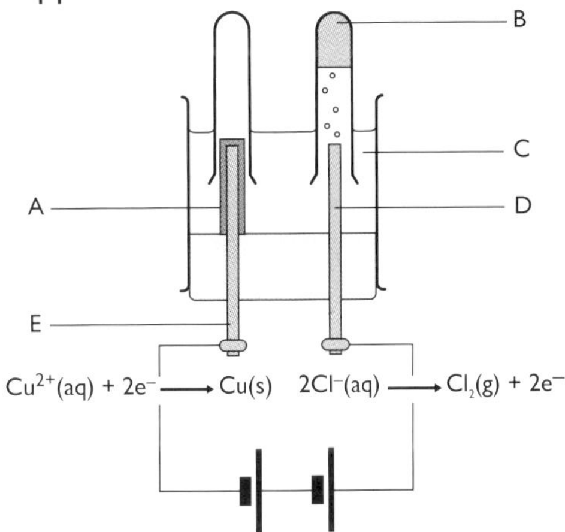

a Name the parts labelled A, B, C, D and E.

b Explain in words what is happening at the two electrodes.

c Give one industrial use of this type of electrolysis.

34 Electrolysis calculations

Use the data tables on page 60 to help you answer these questions.

1 When liquid sodium chloride is electrolysed, the half-equations are:
$Na^+ (l) + e^- \rightarrow Na(l)$
$Cl^-(l) \rightarrow Cl_2(g) + e^-$

- **a** Copy these half equations and balance them so they involve the same number of electrons.
- **b** If an electrolysis cell produces 12.5 g of sodium metal, how many grams of chlorine gas will be produced?

2 A solution of copper bromide is electrolysed. After passing electricity through the solution, it is found that 5.08 g of copper and 12.8 g of bromine have been produced.

- **a** Calculate the empirical formula of copper bromide.
- **b** Write the symbols for a copper ion and a bromide ion.
- **c** Write half equations for the reactions going on at the cathode and anode.

3
- **a** What is a faraday?
- **b** What is a coulomb?
- **c** How many coulombs in a faraday?

4 Complete and balance the following half equations for reactions at electrodes during electrolysis.

- **a** $Cl^- - e^- \rightarrow Cl_2$
- **b** $Cu^{2+} + \ldots \rightarrow Cu$
- **c** $O^{2-} - e^- \rightarrow O_2$

5
- **a** Write the half equations for the electrolysis of aluminium oxide.
- **b** In the electrolysis of aluminium oxide, 54 g of molten aluminium is collected at the cathode. What volume of oxygen would you expect to be given off at the anode?

6 Electrolysis can be used to coat one metal with another. For example, if a fork is used as the cathode with a pure silver anode and the electrolyte contains silver ions (e.g. silver nitrate solution) silver is plated onto the fork.

- **a** Produce a diagram to show what is happening during the silver plating of a fork.
- **b** Give half equations for the events at the electrodes when an object is silver plated.

7 Half equations for the electrolysis of copper bromide solutions are:
cathode: $Cu^{2+}(aq) + 2e^- \rightarrow Cu(s)$
anode: $2Br^-(aq) - 2e^- \rightarrow Br_2(aq)$
Use the two half equations to help you calculate the following. 16 g of bromine are produced at the anode when a solution of copper bromide is being electrolysed. How much copper is produced at the cathode ?

8 During the production of aluminium, the two half equations involved are:
cathode: $4Al^{3+}(l) + 12e^- \rightarrow 4Al(l)$
anode: $6O^{2-}(l) - 12e^- \rightarrow 3O_2(g)$
5.4 kg of aluminium is formed. What mass of oxygen would you expect to collect?

9 0.03 faraday of electricity are passed through sodium hydroxide solution using platinum electrodes.

- **a** Write down the reactions taking place at the electrodes.
- **b** Calculate the number of moles of each gas produced and the volume each gas would take up at stp.
- **c** Calculate how long it would take to complete the passage of 0.03 faraday if a current of 2.5 amps was passed through the solution.

35 Metal extraction

1 Copy and complete these sentences. Use the words below to fill in the gaps.

carbon concentrating Earth hydrogen ore reducing

Metals or metal compounds are usually found in the crust of the .... mixed with other substances. If there is enough metal or metal compound in a rock to be worth extracting, then the rock is called an .... Extracting a metal from its ore usually involves two steps .... and then smelting (.............................). Two substances that can be used in the last stage are .... and ....

2 Old-time gold prospectors used panning to separate gold from surrounding material. Use words and diagrams to explain how panning works.

3 Zinc can be found in the Earth's crust combined with sulphur as zinc sulphide. To produce the pure metal, the sulphide must undergo a heating stage followed by a reduction stage.

a Write a word equation to describe what happens when zinc sulphide is heated.

b In the second stage of the process we could use either hydrogen and carbon.

  i Look at the reactivity series on page 60 and decide which of these could be used to produce zinc.

  ii Explain your answer using a word equation.

4 Rearrange these sentences to describe the production of aluminium from bauxite. Copy them out in the correct order.

- Aluminium metal is formed at the cathode.
- Aluminium is extracted by electrolysis, since it is above carbon in the reactivity series.
- A current passes through carbon electrodes into the molten mixture.
- Aluminium is found in the ore bauxite.
- Oxygen forms at the other electrode, reacting with it to form carbon dioxide.
- The bauxite is dissolved in cryolite, and heated to almost 1000 °C.

5 **a** In the production of aluminium from its ore, why is it not possible to use copper electrodes?

b Which metals could be used to make the electrodes? (Use the data table on page 60 to help you.)

6 Use the reactivity series of metals (page 60) to answer these questions. Explain the reasons for your answers carefully.

a Which metals must be extracted from their ores using electrolysis?

b Which metals are found naturally in the Earth's crust, uncombined with other elements?

7 There is more aluminium in the Earth's crust than any other metal. So what are the arguments for recycling aluminium?

8 The diagram shows the electrolysis of bauxite (aluminium oxide).

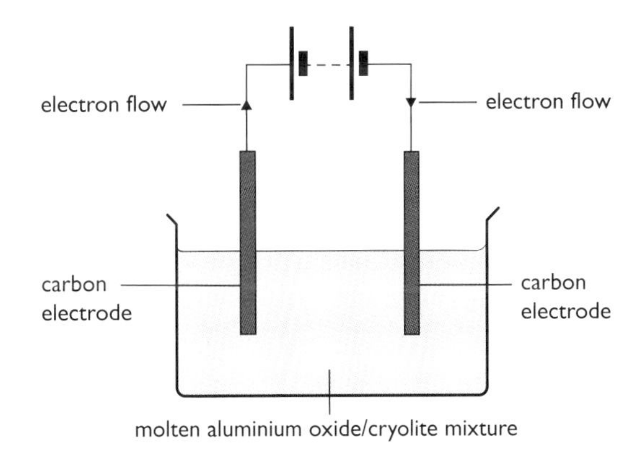

a Write down half-equations for the processes occurring at

  i the anode **ii** the cathode.

b Copy the diagram and add suitable details to show the processes occurring at each electrode.

c Copy and complete the two statements.

  i oxidation (........................ of electrons) occurs at the ....

  ii reduction (........................ of electrons) occurs at the ....

36 The blast furnace

1 Rearrange these sentences to describe the production of iron from iron ore. Copy them out in the correct order.

- Molten iron runs to the bottom of the blast furnace, where it can be tapped off.
- Iron ore is a type of iron oxide.
- Carbon monoxide reduces the iron oxide to iron.
- The slag runs to the bottom of the furnace, where it floats on top of the molten iron.
- In the blast furnace, coke burns in a stream of hot air, and produces a gas called carbon dioxide.
- Carbon dioxide reacts with more carbon to form another gas called carbon monoxide.
- The limestone reacts with acidic impurities to form a molten slag.
- Coke and iron ore are mixed together and fed into a blast furnace together with limestone.
- The reducing agent which is used to remove oxygen from iron oxide is carbon in the form of coke.

2 Copy the diagram of the blast furnace. Use the labels at the top of the next column to replace A–F on your diagram.

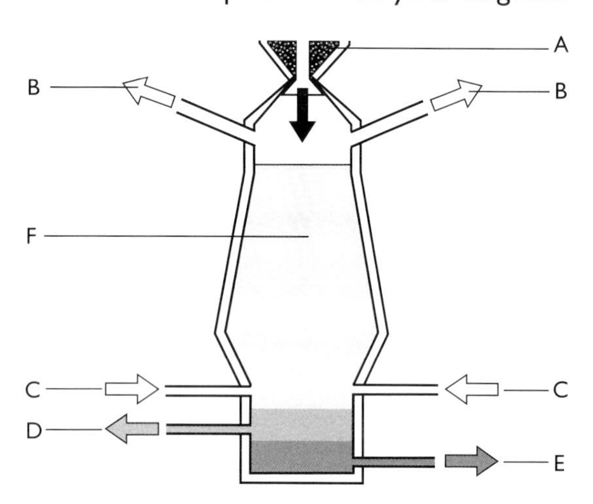

molten iron out **1500 °C**
hot gases out **blast of hot air in**
mixture of iron ore, coke and limestone in **molten slag out**

3 Use the following information to write balanced chemical equations for these reactions which occur in the blast furnace.

a Carbon and oxygen react to form carbon dioxide.

b Carbon dioxide reacts with carbon to form carbon monoxide.

c Iron ore (iron oxide) reacts with carbon monoxide to form iron and carbon dioxide.

d Limestone (calcium carbonate) decomposes to form calcium oxide and carbon dioxide.

e Calcium oxide reacts with impurities such as silica to form slag (calcium silicate).

Chemical formulae you may need:

- carbon dioxide, CO_2
- carbon monoxide, CO
- calcium carbonate, $CaCO_3$
- calcium oxide, CaO
- calcium silicate, $CaSiO_3$
- iron oxide, Fe_2O_3
- carbon, C
- silica, SiO_2
- oxygen, O_2

37 More metals

1. Car body parts are made from iron and steel. Use this to help you explain
   - **a** why cars go rusty
   - **b** why cars rust more rapidly if the paintwork is scratched.

2. Tin is an unreactive metal that is used to cover iron and steel.
   - **a** Explain how this will prevent the iron from rusting.
   - **b** Explain why tin will not be as effective as zinc at protecting the iron, particularly if it gets scratched.

3. The table shows the composition of two different steels. Use the data to help you answer these questions.

Property	Steel 1	Steel 2
max % carbon	0.20	0.22
max % silicon	–	0.55
max % manganese	1.50	1.60
max % sulphur	0.035	0.035
max % phosphorus	0.035	0.035
tensile strength (N/mm^2)	410 – 560	490 – 630
yield strength (N/mm^2)	275	355
impact value (10 × 10: specimen at –20 °C)	27	27

   - **a** Produce a bar chart to compare the composition of the two steels excluding iron.
   - **b** If the maximum percentage of the other ingredients is present, what is the percentage of iron present in each type of steel?

4. **a** Display the data from the table in **Q3** to compare the properties of the two different steels.
   - **b** Which components of the steel do you think are affecting the tensile strength and the yield strength of the alloy? Explain your answer.

5. Explain how a steel kettle might be electroplated with copper to make it more decorative. Use a diagram to help your explanation.

6. **a** What percentage of zinc is extracted by electrolysis?
   - **b** Describe the extraction of zinc by electrolysis, giving the half reaction for the cathode.
   - **c** What is the alternative method by which zinc can be extracted from its ore? Explain how zinc can be extracted in this way.
   - **d** What are the main uses of zinc?

7. **a** What is the thermite process?
   - **b** Give a word equation for the thermite reaction.
   - **c** How is pure chromium used and why?

38 Ordering elements

❶ Copy and complete these sentences. Use the words below to fill the gaps.

different elements repeating similar

When the first 20 .... are listed in order of atomic number, they show a .... pattern of properties. Elements next to one another are often very .... but elements with .... properties appear after a count of 8.

❷ Copy the table of elements.

Atomic no.	Mass no.	Element	Properties
1	1	hydrogen	a very reactive gas
2	4	helium	an inert (unreactive gas)
3	7	lithium	a soft, very reactive metal
4	9	beryllium	a reactive metal
5	11	boron	a solid non-metal
6	12	carbon	a solid non-metal
7	14	nitrogen	a non-metal
8	16	oxygen	a reactive non-metal
9	19	fluorine	a very reactive non-metal (gas)
10	20	neon	?
11	23	sodium	?
12	24	magnesium	a reactive metal
13	27	aluminium	a reactive metal
14	28	silicon	?
15	31	phosphorus	a non-metal
16	32	sulphur	?
17	35.5	chlorine	?
18	40	argon	an inert gas
19	39	potassium	a soft, very reactive metal
20	40	calcium	?

a Use the 'count to 8' rule to fill in the missing properties.

b This pattern is good, but not perfect. What is odd about elements number 5 and 13?

c Originally, the elements were ordered by increasing atomic mass. Which elements would be 'out of position' if you did this?

d When Newlands first noticed this pattern in 1864, this anomaly didn't bother him. Why not? (Clue: the family of unreactive gases was discovered in 1892.)

❸ **a** Copy out this simple version of the Periodic Table. Put the missing element symbols in their correct places. Choose from Ca, He, S, C, F and Mg. Use **Q2** to help you.

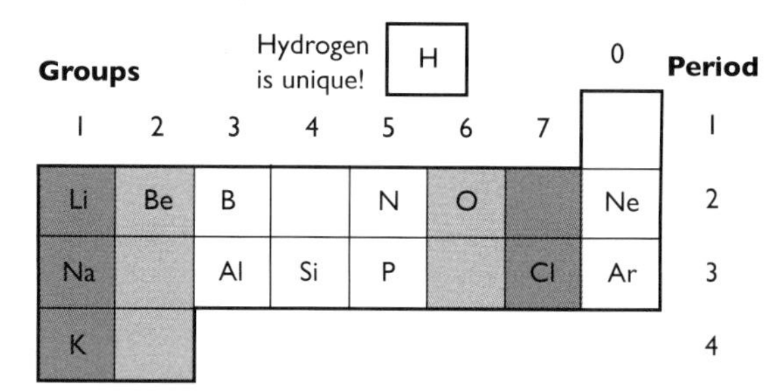

b What do all the elements in

i Group 1 have in common

ii Group 7 have in common?

c What is the name of the element with the symbol

i Na **ii** K?

❹ Redraw your simple version of the Periodic Table from **Q3**. Use the sentences below to annotate it, to explain the link between the structure of the Periodic Table and electron shells.

A The group number tells you how many electrons there are in the outer shell.

B Helium is usually put in Group 0 with the noble gases as its two electrons still give it a full outer shell.

C The period number of an element tells you how many electron shells it has.

D The first period only has 2 elements because the first shell can only take 2 electrons.

E The higher periods have 8 elements, because these shells can take 8 electrons.

39 The Periodic Table

1 The table below gives the melting points (in °C) of the first 20 elements.

Atomic number	Element	Melting point (°C)
1	H	−259
2	He	−272
3	Li	181
4	Be	1278
5	B	2300
6	C	3652
7	N	−210
8	O	−218
9	F	−220
10	Ne	−248
11	Na	98
12	Mg	649
13	Al	660
14	Si	1410
15	P	44
16	S	119
17	Cl	−101
18	Ar	−189
19	K	63
20	Ca	839

a Plot a graph of melting point against atomic number.

b How many of these elements are solid at room temperature (25 °C)?

You will need to use the table in **Q1** and the Periodic Table (page 59) to answer **Q2** and **Q3**.

2 **a** To what group do elements number 2, 10 and 18 belong?

b Are these elements metals or non-metals?

c What do you notice about the melting points of elements 2, 10 and 18?

d To what group do elements number 6 and 14 belong?

e Are these elements metals or non-metals?

f What do you notice about their melting points?

3 **a** To what group do elements number 3, 11 and 19 belong?

b Are these elements metals or non-metals?

c Plot a graph of the melting points of these elements. What happens as the proton number goes up?

d These three elements get more reactive as the proton number increases. The next element in this group is rubidium (Rb, atomic number = 39). Predict its melting point and suggest how reactive it will be.

4 Copy and complete these sentences. Use the words below to fill the gaps.

atoms **gases** **unreactive**

Group 0 contains the .... noble gases. These colourless .... are made from single .... only.

5 Element Q is a colourless gas that boils at −108 °C. It is very unreactive.

a To which group does Q belong?

b How many electrons must Q have in its outer shell?

c Is the gas made of individual atoms or molecules?

40 Metals

❶ Copy and complete these sentences. Use the words below to fill in the gaps.

air bases ores properties reactive salt water

Metals have many .... that make them useful. Many metals react with other substances such as .... and .... Because of this, most metals are found combined with other elements as ....The method used to extract the metal depends on how .... it is. Some metal compounds react with acids – they are called ....When an acid reacts with a base, a .... is formed.

❷ Copy and complete the table to make a list of places that metals are used around the home.

Metal	Why and where used

❸ Copy and complete each sentence using the correct ending from below.

a Iron can be separated from other metals …

b Choosing the right metal for a job involves …

c Metals can be mixed together …

d Chromium can be added to iron …

e The thermal conductivity of a metal measures how good it is at …

Choose endings from

- to form alloys.
- forming stainless steel.
- using a magnet.
- conducting heat.
- finding out its properties and its cost.

❹ The table shows the properties of some metals.
(The tensile strength of a material is a measure of how strong it is under tension, when two ends of a piece of it are pulled.)

Metal	Density (g/cm^{-3})	Melting point (°C)	Tensile strength
aluminium	2.7	660	70
copper	8.9	1084	130
gold	18.9	1064	78
iron	7.9	1540	211
lead	11.3	327	16
mercury	13.6	–39	–
sodium	0.97	98	low
tungsten	19.4	3410	411

a Which metal is

  i the most dense

  ii the least dense

  iii the strongest?

b Which metal is a liquid at room temperature?

c Why is tungsten used as the filament in electric light bulbs?

d Miniature figures (such as toy soldiers) used to be made from lead.

  i Why?

  ii Explain why this is no longer done.

e A lump of gold is dropped into some mercury. Will it float or sink? Explain your answer.

41 Group 1 and 2 metals

1 Copy and complete these sentences. Use the words below to fill the gaps.

electricity shiny soft alkali low reactive

The metals of Group 1 in the Periodic Table are called the .... metals. They are a family of very .... metals. They tarnish rapidly in air, but are .... when fresh. They conduct heat and .... well but are ...., have low densities and .... melting and boiling points.

2 Lithium (Li), sodium (Na) and potassium (K) are the first three alkali metals, in order down Group 1.

a The balanced symbol equation for burning lithium in air is:
$4Li(s) + O_2(g) \rightarrow 2Li_2O(s)$
Write this as a word equation.

b The alkali metals get more reactive down the group. Copy and complete the table using the phrases below.

Metal	Reaction with water
lithium	**i**
sodium	**ii**
potassium	**iii**

- melts, whizzes around and gas catches fire
- fizzes steadily
- melts and whizzes around

3 Sodium reacts with water
sodium + water → sodium hydroxide + hydrogen

a What would happen if you put pH paper in the water after this reaction? Explain why.

b What is the name given to Group 1? Why do you think it is called this?

c Lithium fizzes steadily in water. Which gas is given off?

d What is the name of the alkali that forms during the reaction in part **c**?

4 **a** What is the common name of the Group 2 elements?

b What does magnesium look like?

5 Both quicklime (CaO) and slaked lime $Ca(OH)_2$ are used to control soil pH.

a Write down the equation for the reaction between quicklime and hydrochloric acid, and for slaked lime and hydrochloric acid

b If 56 g of quicklime are need to neutralise some soil, how much slaked lime would be needed to do the same job?

6 Rearrange the sentences to explain why the Group 1 and 2 metals are more reactive down each group. Copy them out in the correct order.

- The further from the nucleus, the weaker the force holding the electron in place.
- Group 1 and 2 metals lose their outer electron to form positive ions when they react.
- The bigger the atom, the further the outer electron is from the positive nucleus.
- The weaker the force, the 'looser' the outer electron.
- Atoms with 'looser' outer electrons are more reactive.
- The atoms get bigger down the group.

42 Halogens

1 Copy and complete these sentences. Use the words below to fill the gaps.

halogens reactive molecules ionic

Group 7 of the Periodic Table contains a family of very .... non-metals called the .... They form diatomic .... and have coloured vapours. They form .... compounds with metals.

2 Copy and complete the table to show the properties of the halogens, using the words and symbols below. One line in the table has been completed for you.

Element	Symbol	State at room temperature	Colour
fluorine	F	gas	yellow/green
chlorine			
bromine			
iodine			

Br Cl I liquid gas solid green purple brown

3 Copy and complete these sentences. Use the names of the halogens below to fill the gaps.

fluorine (F) chlorine (Cl) bromine (Br) iodine (I)

a The dark, almost black, crystals of .... give off a purple vapour which is used to make fingerprints show up on paper.

b .... was used as a poison gas in World War I, but is now used to kill microorganisms in swimming pools.

c The only two elements which are liquid at room temperature are mercury and ....

d Compounds of .... such as sodium fluoride are sometimes added to drinking water as they help strengthen teeth.

e A brown solution of .... in alcohol was used as an antiseptic for cuts and bruises.

4 The diagrams show how chlorine combines by covalent bonding with carbon, and by ionic bonding with sodium. Draw similar diagrams and write similar equations for the reaction between bromine and carbon, and between bromine and sodium.

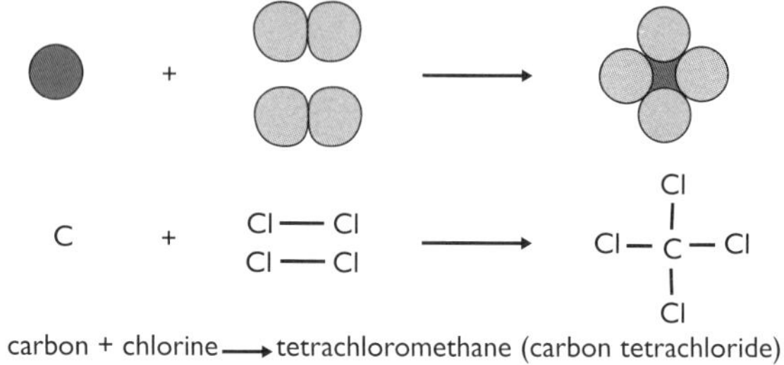

carbon + chlorine → tetrachloromethane (carbon tetrachloride)

$C + 2Cl_2 \rightarrow CCl_4$

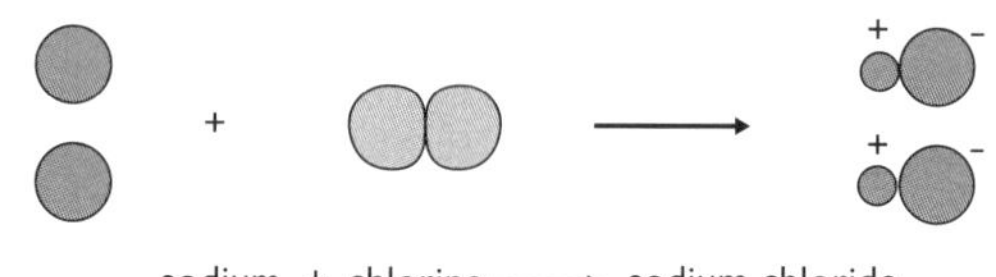

sodium + chlorine → sodium chloride

$2Na + Cl_2 \rightarrow 2NaCl$

5 Rearrange the sentences to explain why the halogens become more reactive up the group. Copy them out in the correct order.

- The closer to the nucleus, the stronger the force that might be able to capture a 'spare' electron.
- The smaller the atom, the closer any 'spare' electrons can get to the positive nucleus.
- Smaller atoms toward the top of the group will find it easier to capture an electron, and so are more reactive.
- Halogens have to gain an outer electron to form negative ions when they react.
- The atoms get smaller as you go up the group.

43 More on halogens

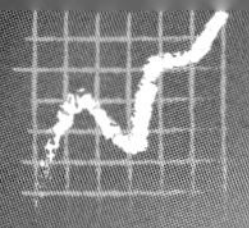

1 Chlorine reacts explosively with hydrogen in sunlight to form hydrogen chloride. This gas is very soluble in water.

a What would happen if you put pH paper in a solution of hydrogen chloride in water?

b What name is given to this solution?

c What compound would form from bromine and hydrogen?

d Would you expect this reaction to be faster or slower than with chlorine? Explain your answer.

2 For many years, compounds of silver and halogens, silver halides, were used in black-and-white photography. For example, silver bromide breaks downs when exposed to light.

$$2AgBr \xrightarrow{\textit{light}} 2Ag + Br_2$$

a Write this reaction out as a word equation.

b Rearrange these sentences to explain how the process works. Copy the sentences out in the correct order.

- Where light falls, the compound breaks up and black grains of silver are formed.
- In a camera, light falls on to the film for a short time.
- This produces a negative image (black for white).
- Where no light falls, the film stays clear.
- Photographic film contains colourless silver bromide.

3 Hydrogen chloride forms when hydrogen gas burns in chlorine.

$H_2 + Cl_2 \rightarrow 2HCl$

a Write this out as a word equation.

b When hydrogen chloride dissolves in water the solution turns pH paper red. What does this tell you?

4 a Draw a table to compare the properties of the halogens fluorine to iodine. Include in your table the state at room temperature, the melting and boiling points and the appearance of the elements.

b Draw a graph to display the melting and boiling point data in your table.

c Explain clearly the way in which the reactions of the halogens with metals differ from the reactions of the halogens with non-metals.

5 Give a word equation and a balanced chemical equation for these reactions.

a fluorine and hydrogen

b iodine and hydrogen

c Describe the conditions under which each of the reactions will take place.

6 a Explain why chlorine will displace bromine and iodine from solutions of their salts.

b Give word and balanced chemical equations for the reactions between

i bromine and potassium iodide

ii chlorine and magnesium bromide.

7 a Give balanced equations for the reactions between

i silver nitrate solution and potassium bromide

ii silver nitrate solution and sodium iodide.

b Explain how these reactions can be used to identify an unknown solution.

8 a Explain the main trends in the reactivity of the halogens down the group. Diagrams will make the explanation clearer.

b Predict the reactions of astatine (At) with

i hydrogen **ii** potassium iodide.

44 Transition metals

1 Copy and complete the following sentences. Use the words below to fill the gaps.

melting transition harder less higher heat

The .... metals are the family of 'everyday' metals. They are shiny and they conduct .... and electricity. They are .... reactive than the alkali metals, but are .... and have higher .... and boiling points and .... densities.

2 Copy and complete the following sentences. Use the names below to fill the gaps.

iron (Fe) nickel (Ni) zinc (Zn)

a Steel is a form of .... that is very strong and is used for building bridges, cars and machinery.
b Copper is alloyed with .... to make brass and with .... to make coins.
c Transition metals are often used as catalyst to speed up reactions. .... is used to make ammonia for fertilisers.

3 **a** Which name is given to the block of metals that wedge in between calcium (20) and gallium (31)? (Clue: unscramble nottiransi).
b The properties of the elements usually change dramatically as you move along the same period. What is unusual about this block of metals?
c Iron is a hard, magnetic metal with a high melting point, which forms coloured compounds. Predict the properties of cobalt (Co).

4 Copy and complete the table at the top right, using the phrases below it.

Property	Alkali metals	Transition metals
reactivity	**a**	**b**
density	**c**	**d**
melting and boiling point	**e**	**f**
colour of salts	**g**	**h**

- very reactive
- less reactive
- high
- sink in water
- can float on water
- colourless
- often coloured
- low

5 **a** Compare the way in which iron forms iron oxide to the formation of black copper oxide.
b Explain the impact this has on the way they are used.
c What else affects their usefulness?

6 Copper is sometimes used to cover the roofs of buildings and for the outer skin of statues (for example, the Statue of Liberty in New York harbour). With time the copper turns a green colour. In industrial areas, this green colour is usually due to a mixture of copper sulphate and copper hydroxide. In other areas, it is usually a mixture of copper carbonate and copper hydroxide.
a Explain why copper sulphate is formed in industrial areas but not elsewhere.
b Where does the carbon come from to form copper carbonate?
c Suggest word equations to describe the formation of the green colour on the roofs of buildings covered with copper.

7 Give word and chemical equations for the reaction between
a copper oxide and sulphuric acid
b copper sulphate and ammonia solution.
c What colour is the complex copper ion formed in this reaction?

45 Combustion

1 Copy and complete these sentences. Use the words below to fill the gaps.
oxides oxygen reacting combustion nitrogen

The air is roughly four-fifths .... and one-fifth .... When things burn in air, they are .... with the oxygen. New compounds called .... are formed, and energy is given out. The scientific word for burning is ....

2 When carbon burns in air it forms the colourless gas, carbon dioxide.

a Copy and complete the word equation for this reaction.

carbon + ... → ....

b When carbon burns, it seems to disappear. Where does it go?

3 When hydrogen burns, it forms hydrogen oxide vapour.

a What is the common name we give to this new chemical? (Hint: it's usually a liquid when you see it).

b Copy and complete the word equation for this reaction.

hydrogen + ... → ....

4 Fuels are chemicals which we burn to get energy.

a Copy and complete the generalised word equation for this reaction.

fuel + ... → waste gases + energy

b What type of chemicals will these waste gases be? (Check back to **Q1**.)

5 Many fuels, such as oil, are chemicals that contain both carbon and hydrogen.

a Which two oxides would you expect to find in the waste gases when this type of fuel burns?

b Methane (CH_4) Is a gas which is often used as a fuel. Write the word and chemical equation for the reaction of oxygen with methane.

6 This apparatus is used to identify the waste gases produced when a candle burns.

a Copy the diagram. Use the labels below in place of A–F on your diagram.

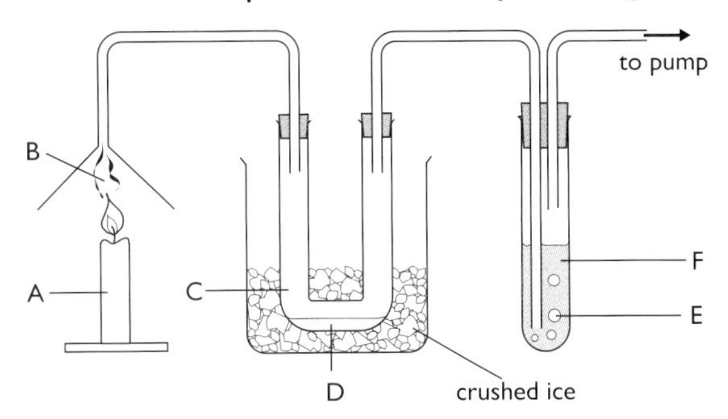

- The crushed ice cools the gases.
- The limewater turns milky.
- Candle wax contains carbon and hydrogen.
- The gas bubbles through limewater.
- The waste gases rise into the funnel.
- A colourless liquid condenses and collects.

b The colourless liquid turns white anhydrous copper sulphate powder blue. What is this liquid? Label this on your diagram.

7 Special blowtorches can weld or cut metal under water. They mix gases from two separate cylinders. The two gases then burn together, even under water. One of the gases is a fuel. What must the other gas be? Explain your answer.

8 **a** If you heat potassium nitrate it melts and then bubbles as a gas is given off. The gas relights a glowing splint. What do you think the gas might be?

b Gunpowder contains carbon and sulphur, mixed with potassium nitrate. Gunpowder will burn, even if there is no air. Explain why this is possible.

c Which two waste gases do you think will be formed when gunpowder burns? Explain your answer.

46 Gases from the air

1 Copy and complete these sentences. Use the words below to fill the gaps.

acidic dioxide rain fish sulphuric sulphur

Many fuels contain .... as an impurity. When this burns it produces sulphur .... gas. This gas dissolves in the .... and reacts with the air to form weak .... acid. This makes the rain .... Acid rain can damage buildings and kill .... and plants.

2 These pie charts show the sources of three pollutant gases in the atmosphere (of Canada).

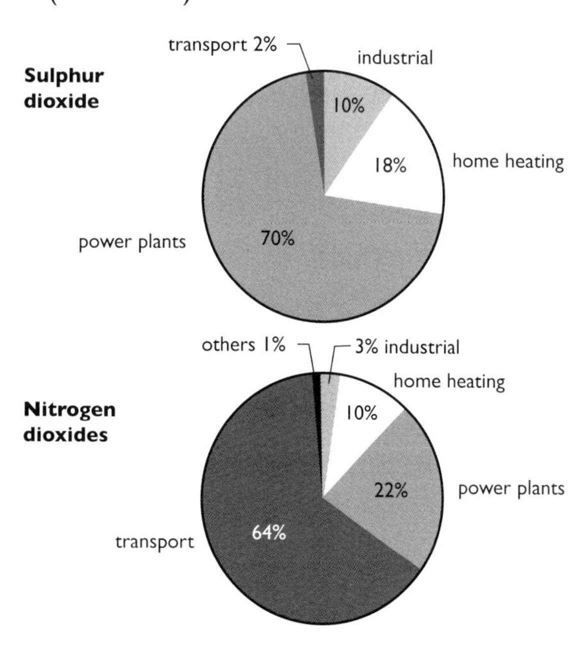

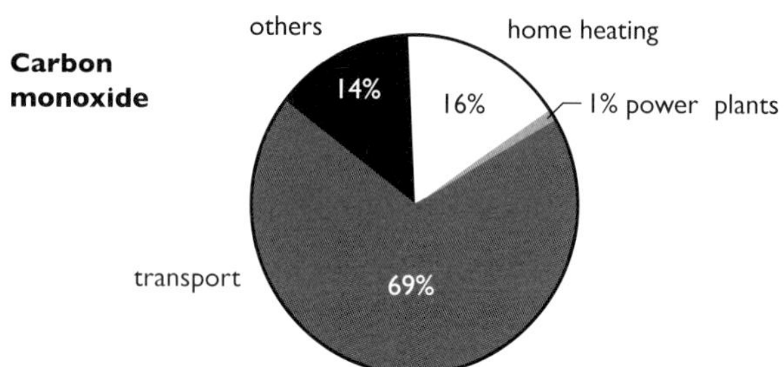

a What is the largest source of sulphur dioxide?

b What are the major pollutants produced by the transport industry?

c Nitrogen oxides and carbon monoxide combine to produce photochemical smog. Why do you think this type of smog is most common in the cities?

3 Summarise the industrial process by which oxygen and nitrogen gases are extracted from the air.

4 **a** What are the main properties of carbon dioxide?

b What are the main uses of carbon dioxide?

c How is the reaction of carbon dioxide with calcium hydroxide solution used as a test for the presence of the gas?

5 **a** Draw and annotate a diagram to show how ammonia can be made in the laboratory.

b Why is the test tube used to collect the ammonia upside down?

c Why is damp red litmus held at the exit of the test tube?

47 Analysis

1 **a** What is a flame test?

b Describe how you would carry out a flame test.

c Copy and complete this table.

Element	Flame colour
lithium	**a**
b	golden yellow
c	lilac
calcium	**d**

2 Copy and complete this table.

Add sodium hydroxide solution	Flame test	Metal ion
nothing observed	lilac	**a**
white precipitate	brick red	**b**
c	**d**	Fe^{3+}
white precipitate which dissolves as more sodium hydroxide solution added	nothing observed	**e**
light green precipitate which slowly turns reddish brown	nothing observed	**f**
g	**h**	Na^{+}

3 Four samples of chemicals have been found by detectives hunting a gang of international chemical thieves. Which of the four powders can be positively identified? Give your reasons.

Powder	Appearance	Add dilute acid	Effect of heat
A	green	fizzes	green powder turns black and a gas is evolved which turns limewater milky
B	white	nothing	nothing
C	white	fizzes	white powder turns yellow while hot cooling to white again, and a gas is evolved which turns limewater milky
D	green	fizzes	nothing

4 Compound A is a white solid which dissolves in water to produce a colourless solution. When this solution is acidified with nitric acid and silver nitrate is added, a white precipitate is produced. A flame test of A produces a bright red flame. Deduce the name of compound A and give your reasoning.

5 Compound B is a white solid which dissolves in water to give a blue solution. When this solution is acidified with hydrochloric acid and barium chloride is added, a white precipitate is produced. Deduce the name of compound B and give your reasoning.

6 Copy and complete this table.

Add dilute acid	Add sodium hydroxide solution and warm	Add sodium hydroxide solution and warm – then add aluminium	Flame test	Substance
nothing observed	nothing observed	gas evolved, turns damp red litmus blue after aluminium added	golden yellow	**a**
fizzing, gas turns limewater milky	gas evolved turns damp red litmus blue	nothing observed	nothing observed	**b**
c	**d**	**e**	**f**	calcium carbonate

7 Copy and complete this table.

Add nitric acid and then silver nitrate solution	Flame test	Substance
creamy precipitate	golden yellow	**a**
white precipitate	green	**b**
yellow precipitate	lilac	**c**

8 **a** What is ammonia?

b How would you test for ammonia and the ammonium ion?

9 How would you test for nitrate ions in an unknown solution?

48 Crude oil

❶ Copy and complete these sentences. Use the words below to fill the gaps.
atoms   hydrocarbons   molecules

Crude oil is a mixture of compounds called .... These contain particles called .... made from carbon and hydrogen .... joined together in chains.

❷ Copy the diagram that shows the apparatus used to separate crude oil into its fractions in the laboratory. Use the labels below in place if the letters A–E on your diagram.

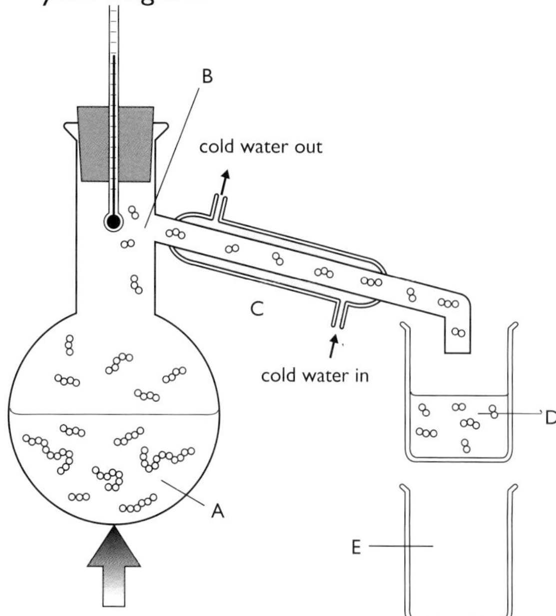

- The lighter, smaller particles boil off first, at a low temperature.
- Cold water cools the gas particles and makes them condense.
- Crude oil is heated in the flask.
- The beaker is changed as the temperature rises, to collect different fractions.
- The liquid collects in the beaker.

❸ **a** Copy the diagram of a fractionating column. Use the labels below the diagram in place of A–E on your diagram.

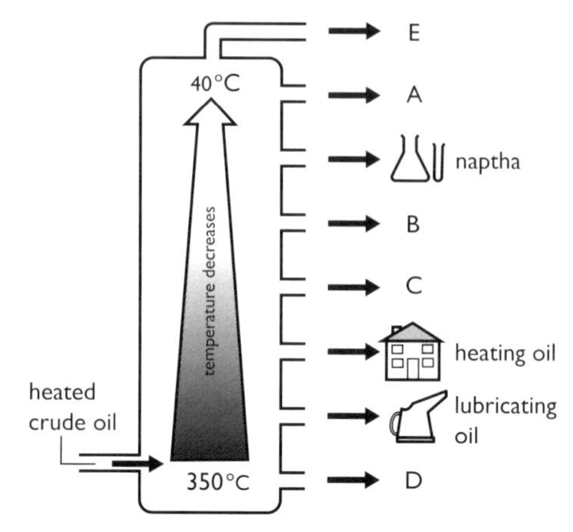

- 15–30% gasoline (petrol) is used as fuel for cars (b.p. 40–100 °C)
- 40–50% bitumen is used to make roads (b.p. over 400 °C)
- 15–20% diesel is used as lorry fuel (b.p. 220–300 °C)
- 10–15% paraffin (kerosene) is used for jet fuel (b.p. 150–240 °C)
- 1–2% petroleum gases used for fuel (b.p. 40 °C)

b Which other substances would you expect as products of the fractional distillation of crude oil?

❹ The table shows information about some hydrocarbons.

Number of carbon atoms	Boiling point (°C)
5	36
6	69
7	99
9	151
10	174

a Plot a graph of boiling point against number of carbon atoms.
b What happens to the boiling point as the number of carbon atoms goes up?
c Use your graph to predict the boiling point of octane – a hydrocarbon with eight carbon atoms in its chain.
d In a laboratory experiment, Sasha collected three fractions from a sample of crude oil. They boiled at
A 50–90 °C   B 100–140 °C
C 150–200 °C
Which of these could be used as petrol? Explain your answer.

49 Organic chemistry

1. What is the difference between organic and inorganic compounds?

2. **a** What is an alkane?
   b Draw the structure of a molecule of pentane.
   c How do alkanes react in a plentiful supply of oxygen?
   d Give two uses of members of the alkane family.
   e Write a balanced equation showing how butane burns in air.

3. **a** What is an alkene?
   b Draw the structure of a molecule of pentene.
   c How are the alkenes formed from the alkanes?
   d Give two reasons why the alkenes are not used as fuels.
   e Write a balanced equation showing how butane burns in air.

4. Give the structural and molecular formula for an alkane containing
   a three carbon atoms
   b five carbon atoms
   c eight carbon atoms.

5. Give the structural and molecular formula for an alkene containing
   a three carbon atoms
   b five carbon atoms
   c eight carbon atoms.

6. Use the general formula of the alkanes to work out how many hydrogen atoms there are in an alkane with these numbers of carbon atoms.
   a 6 **b** 10 **c** 7 **d** 15

7. Use the general formula of the alkenes to work out how many hydrogen atoms there are in an alkene with these numbers of carbon atoms.
   a 6 **b** 12 **c** 20 **d** 17

8. Explain in terms of the structure of the molecule why butane is used as a fuel and butene is not.

9. Use the data from this table to answer the questions below.

Isomer	pentane	2-metylbutane	2,2-dimethylpropane
Boiling point	36.3 °C	28 °C	10 °C

   a Draw a bar chart showing the boiling points of each of the three isomers of pentane.
   b What causes the differences in the boiling points of these three compounds which all have the same molecular formula, C_5H_{12}?

10. **a** Why is an unsaturated hydrocarbon more reactive than a saturated one?
    b Explain why an alkene but not an alkane turns bromine water colourless.

11. **a** What is an isomer?
    b Draw at least three isomers for hexane C_6H_{14}.
    c How many isomers can you draw for octane, C_8H_{18}?
    d Take the isomers you have worked out in part **c** and name each one systematically.

50 New products from crude oil

1 Copy and complete these sentences. Use the words to fill the gaps.

cracking plastics temperatures hydrocarbon

The fractional distillation of crude oil produces more large .... molecules than are needed. Fortunately, these can be 'chopped up' into smaller molecules by a process called .... This happens when the hydrocarbons are heated to very high .... Cracking makes more petrol, but it also makes special molecules that are used to make ....

2 Draw a pie chart to show the relative proportions of different products that may be obtained from crude oil if 1000 000 tonnes makes

- 30 000 tonnes petrol
- 7 000 tonnes naphtha
- 10 000 tonnes kerosene
- 30 000 tonnes diesel
- 20 000 tonnes fuel oil
- 3 000 tonnes other.

3 Copy and complete these sentences. Choose the correct word from each pair.

Crude oil hydrocarbons are called **alkanes/alkenes**. Their carbon atoms are joined by **single/double** bonds. They cannot form any extra bonds so they are said to be **saturated/unsaturated**. When hydrocarbons are cracked, **alkanes/alkenes** such as ethene are formed. Ethene has a **single/double** bond. This can open up to add more atoms, so ethene is said to be **saturated/unsaturated**.

4 **a** Draw a diagram of the simple alkane, hexane, which has six carbon atoms (C_6H_{14}).

b Hexane can be cracked to give the alkene ethene (C_2H_4) as well as an alkane **A**. Draw a diagram of ethene.

c In alkane **A**, how many atoms must there be of

i carbon atoms

ii hydrogen atoms?

d Draw a diagram of alkane **A**.

5 **a** Why do the fractions of crude oil from fractional distillation need more refining?

b What happens in a catalytic cracker?

c Why are small molecules sometimes put into a catalytic cracker?

51 Polymers

1 **a** List as many uses for oil-based fuels as you can.
b Why are alternatives to oil-based fuels being investigated?

2 **a** What is a plastic?
b Why are plastics so useful?
c What is the effect of using different monomer units in a polymer?

3 Produce a leaflet which explains to the public why plastics can be a threat to the environment and encourages them to recycle plastics and buy biodegradable products when they can.

4 **a** Why are some large molecules simply very big molecules when others are described as polymers?
b Why are alkenes often involved in the formation of polymers?

5 **a** What is meant by the term addition polymerisation?
b How does addition polymerisation differ from other types of polymerisation?
c How does ethene react to form the polymer polythene?

6 **a** Use the diagrams top right to help you explain how ethene polymerises to make polythene.

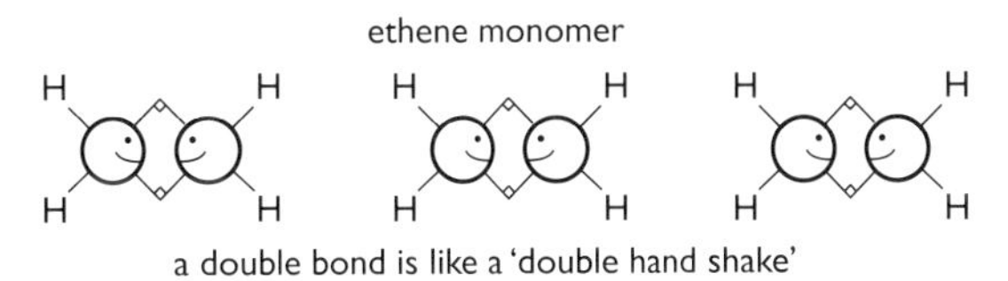

a double bond is like a 'double hand shake'

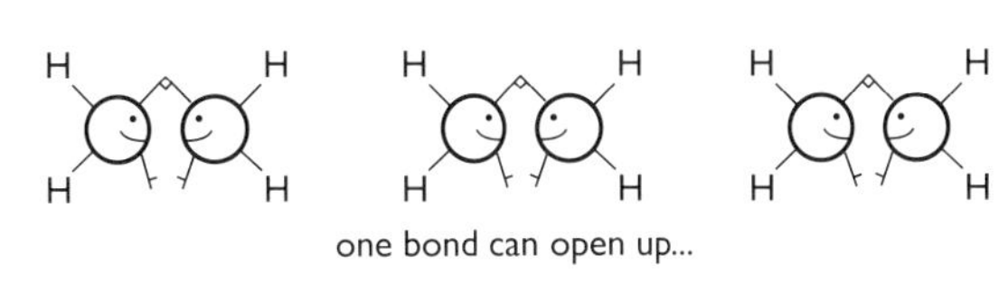
one bond can open up...

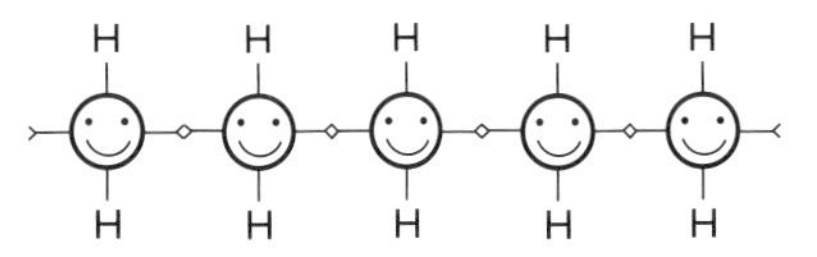
and the monomers can link up in an addition polymer

polymerisation

the double bond opens

...to give a repeating unit with a free 'arm' at each end

b Propene is an alkene with three carbon atoms. It is like ethene, with an extra CH_3 in place of one of the hydrogen atoms. It can be drawn like this.

or

Draw diagrams like those for ethene, to show how propene could undergo addition polymerisation to form polypropene.

7 Polybutene is formed from monomer butene units. Draw a diagram and for this reaction.

52 Ethanol

1 Give three common uses for ethanol, (C_2H_5OH) giving a specific example in each case.

2 Why is it so important that the fermentation of sugar takes place without oxygen?

3 Draw a flow chart to summarise the stages in the production of pure ethanol from sugar.

4 **a** Give the molecular and structural formula for butanol and pentanol.
b The higher alcohols have isomers. Give the possible isomers of butanol and propanol.

5 Using the information on these pages and other resources, make a presentation on the use of ethanol as a fuel for the future, explaining the advantages and the difficulties involved.

6 **a** Why is it so important that the cork in a bottle of wine should be completely airtight?
b How would you expect wine vinegar to be made?

7 **a** What would you expect to happen if propanol (C_3H_7OH) was left exposed to the air?
b Give a word equation to show the reaction.

8 Give the molecular and structural formula of propanoic acid and pentanoic acid.

9 **a** Give an example of a reaction between a carboxylic acid and an alcohol.
b Which catalyst is usually needed for this reaction to take place?
c What are the products of this type of reaction usually used for?

10 **a** Draw up a table to evaluate the two different methods of producing ethanol to show:
- rate of reaction
- quality of the product
- use of finite resources
- batch process or continuous process
- reaction conditions.

b Explain which process you would support for future development and why.

11 Give molecular and structural equations for the reaction between propanol and propionic acid.

12 **a** Give the molecular and structural equations for the reactions of methanol with sodium.
b Which of these alcohols would you expect to react most vigorously with sodium: methanol, ethanol or propanol? Explain your answer.

Periodic Table

Period	1	2						Group					3	4	5	6	7	0
1								1 H Hydrogen 1										4 He Helium 2
2	7 Li Lithium 3	9 Be Beryllium 4											11 B Boron 5	12 C Carbon 6	14 N Nitrogen 7	16 O Oxygen 8	19 F Fluorine 9	20 Ne Neon 10
3	23 Na Sodium 11	24 Mg Magnesium 12											27 Al Aluminium 13	28 Si Silicon 14	31 P Phosphorus 15	32 S Sulphur 16	35.5 Cl Chlorine 17	40 Ar Argon 18
4	39 K Potassium 19	40 Ca Calcium 20	45 Sc Scandium 21	48 Ti Titanium 22	51 V Vanadium 23	52 Cr Chromium 24	55 Mn Manganese 25	56 Fe Iron 26	59 Co Cobalt 27	59 Ni Nickel 28	63.5 Cu Copper 29	65 Zn Zinc 30	70 Ga Gallium 31	73 Ge Germanium 32	75 As Arsenic 33	79 Se Selenium 34	80 Br Bromine 35	84 Kr Krypton 36
5	86 Rb Rubidium 37	88 Sr Strontium 38	89 Y Yttrium 39	91 Zr Zirconium 40	93 Nb Niobium 41	96 Mo Molybdenum 42	99 Tc Technetium 43	101 Ru Ruthenium 44	103 Rh Rhodium 45	106 Pd Palladium 46	108 Ag Silver 47	112 Cd Cadmium 48	115 In Indium 49	119 Sn Tin 50	122 Sb Antimony 51	128 Te Tellurium 52	127 I Iodine 53	131 Xe Xenon 54
6	133 Cs Caesium 55	137 Ba Barium 56	139 La Lanthanum 57	179 Hf Hafnium 72	181 Ta Tantalum 73	184 W Tungsten 74	186 Re Rhenium 75	190 Os Osmium 76	192 Ir Iridium 77	195 Pt Platinum 78	197 Au Gold 79	201 Hg Mercury 80	204 Tl Thallium 81	207 Pb Lead 82	209 Bi Bismuth 83	210 Po Polonium 84	210 At Astatine 85	222 Rn Radon 86
7	223 Fr Francium 87	226 Ra Radium 88	227 Ac Actinium 89															

Key

Key
Relative atomic mass
Symbol
Name
Atomic number

Data Tables

Elements, their symbols and relative atomic mass

Element	Symbol	Relative atomic mass (A_r)
aluminium	Al	27
bromine	Br	80
calcium	Ca	40
carbon	C	12
chlorine	Cl	35.5
copper	Cu	63.5
fluorine	F	19
hydrogen	H	1
iron	Fe	56
magnesium	Mg	24
nitrogen	N	14
oxygen	O	16
phosphorus	P	31
potassium	K	39
silicon	Si	28
sodium	Na	23
sulphur	S	32
vanadium	V	51
zinc	Zn	65

Reactivity series of some elements

potassium	most reactive
sodium	
lithium	
calcium	
magnesium	
aluminium	
(carbon)	
zinc	
iron	
(hydrogen)	
copper	least reactive

Energies of some chemical bonds

Bond	Bond energy kJ/mole	Bond	Bond energy kJ/mole
Br–Br	+193	H–Br	+366
C–C	+347	H–O	+463
C–O	+358	H–I	+298
C–H	+413	H–H	+436
C–N	+286	H–Cl	+431
C–F	+467	O=O	+496
C–Cl	+346	N–N	+944
C=O	+743	N–H	+388
C–I	+234	I–I	+151
Cl–Cl	+243		

Selected ions

aluminium	Al^{3+}
sodium	Na^{+}
zinc	Zn^{2+}
iron	Fe^{2+}
silver	Ag^{+}
chlorine	Cl^{-}
fluorine	F^{-}
iodine	I^{-}
oxygen	O^{2-}

Glossary

acid A solution with a pH less than 7.
acid rain Produced when gases such as sulphur dioxide dissolve rainwater.
activation energy The energy which must be supplied to reactants before they will react.
alkali A solution with a pH greater than 7.
alkali metal Elements in Group I of the periodic table.
alkane Hydrocarbon with the general formula C_nH_{2n+2}.
alkene Hydrocarbon containing one carbon–carbon double bond with the general formula C_nH_{2n}.
alloy A compound made from two or more metals.
anode The positive electrode.
atom The tiny particles which make up an element.
atomic number The number of protons in the nucleus of an atom. All atoms of the same element have the same atomic number.
base Substance that neutralises an acid.
bauxite An ore of aluminium, more common than cryolite.
blast furnace The process used for making iron from iron ore, using limestone and coke.
brine Sodium chloride solution.
catalyst A substance that increases the rate of a reaction without altering anything else. It is not used up during the reaction, and can be used over and over again.
cathode The negative electrode.
chemical reaction When two or more substances combine to make one or more new substances.
combustion The process of burning.
compound Two or more chemical elements joined together.
conductor Something that will allow electricity/heat to pass easily through it.
Contact process The industrial process used for making sulphuric acid.
cosmic radiation Radiation from space.
covalent bond Bond between elements in which electrons are shared.
cracking Breaking up large hydrocarbons into smaller ones using heat.
crude oil Natural mixture of many different hydrocarbons.
cryolite An ore of aluminium, less common than bauxite.
ΔH Energy change during a reaction; negative when energy released, positive when energy absorbed.
diffusion Net movement of particles by random motion from an area of higher concentration to one of lower concentration.
displacement reaction A reaction in which one substances replaces another.
distillation Separation of a substance from a mixture based on differing boiling points.
electrode A conductor placed in a solution in order to pass electricity through it.
electrolysis Splitting up a substance using electricity.
electrolytes Solutions that conduct electricity.
electrons Negatively charged particles found in atoms.
element Substances that cannot be broken down into simpler substances, for example carbon and oxygen.
empirical formula Simplest formula for a substance showing the ratio of elements in it.
endothermic reaction A reaction which absorbs energy.
enzymes Biological catalysts.
equilibrium When the rate of a reversible reaction is exactly the same in both directions.
equilibrium mixture The mixture of products and reactants that is present when a reversible reaction reaches equilibrium.
exothermic reaction A reaction which releases energy.
fermentation The conversion of sugar to alcohol and carbon dioxide.

Glossary

fertiliser A substance added to the soil to replace nutrients taken out by plants.
flame test Heating a substance in a flame to identify a metal ion.
fractional distillation Separation of many substances in a mixture by heat, using the fact they have different boiling points.
gas State of a substance where particles move freely and fill the container.
Haber process The process used to make ammonia from nitrogen and hydrogen.
half equation Equation showing only the reaction of the ion at an electrode during electrolysis.
halogens Elements in Group 7 of the periodic table.
hazard symbol A symbol used to identify the hazards associated with particular chemicals.
hydrocarbon Substance made up only of carbon and hydrogen.
indicator A substance used to show whether a solution is acid or alkali.
insulator Something that does not allow electricity/heat to pass easily through it.
intermolecular force Forces that hold molecules together in giant structures.
ion An atom which has lost or gained one or more electrons, becoming charged.
ionic bond Bond between two atoms in which atoms are transferred.
ionic substance A substance made up of a combination of positive and negative ions.
isomers Two or more substances containing chemical elements in the same proportions but with different arrangement of atoms.
litmus An indicator. In acid solutions litmus is red, in alkaline solutions it is blue.
liquid State of a substance with no fixed shape but a definite volume.
mass A measure of the amount of matter in a body. No matter where an object is, its mass is always the same.
mass number The number of protons plus the number of neutrons in the nucleus of an atom.
melt Change state from solid to liquid.
metal A shiny substance thatconducts electricity. All metals except mercury are solid at room temperature.
metallic bond Chemical bond produced by attraction between mobile electrons and fixed positively charged metallic ions.
mixture Combination of two or more elements that are not chemically combined.
mole The mass of substance containing 6.02×10^{23} particles.
molecule A group of atoms joined together by chemical bonds.
monomer The single units from which polymers are made.
net energy transfer Energy released or absorbed during a reaction.
neutral A solution with a pH of exactly 7.
neutralisation The reaction between an acid and a base to produce one or more neutral substances.
neutron Uncharged particle found in the nucleus of an atom.
nitrates The source of nitrogen in the soil for most plants.
non-metals Most non-metal elements have low melting and boiling points, and do not conduct electricity.
ore Rock from which useful quantities of a metal can be extracted.
oxidation The removal of electrons from a substance, or addition of oxygen.
Periodic table Arrangement of elements in order of increasing atomic number, elements with similar properties appearing in the same column.
phenolphthalein An indicator. In acid solutions phenolphthalein is clear, in alkaline solutions it is pink.
polymer Substance made of a repeating sequence of monomers.
polymerisation The formation of a polymer from monomers.
product A substance made in a chemical reaction.
proton Positively charged particle found in the nucleus of an atom.
proton number See atomic number.
reactant A substance at the start of a chemical reaction.
reaction rate The speed of a reaction. The greater the reaction rate, the faster the reaction goes.

Glossary

reactivity Used to describe the chemical behaviour of a substance. Reactive substances join up easily with other substances, unreactive substances do not.

reactivity series A list showing substances in order of reactivity.

reduction The addition of electrons to a substance, or removal of oxygen.

relative atomic mass (A_r) The mass of an atom compared to other atoms. A_r for hydrogen is 1, and A_r for carbon is 12, so a carbon atom is 12 times heavier than a hydrogen atom.

relative formula mass (M_r) The sum of all the relative atomic masses of the atoms making up a compound.

reversible reaction A reaction that can go forwards and backwards.

salt A substance formed when an acid and a base react together.

saturated hydrocarbon Hydrocarbon with no double bonds.

shell The position in which electrons orbit a nucleus of an atom.

slag Waste material produced in the blast furnace during the production of iron.

solid State of matter with fixed volume and shape.

steel A range of metals containing iron and other substances such as carbon, nickel and other elements.

titration Method for finding out exactly how much of two solutions are needed to neutralise each other, using an indicator.

transition metal Elements which sit between Groups 2 and 3 of the periodic table.

unsaturated hydrocarbon Hydrocarbon containing at least one carbon–carbon double bond.

yield The amount of product made in a chemical reaction. It is usually expressed as a percentage of the maximum possible amount of product(s) that can be made from a given quantity of starting materials.

Glossary